14th High-Tech Plasma Processes Conference (HTPP 2016)

Journal of Physics: Conference Series Volume 825

Munchen, Germany
3 July 2016

Editors:

Dirk Uhrlandt
Philippe Teulet
Jochen Schein

ISBN: 978-1-5108-3980-9
ISSN: 1742-6588

Printed from e-media with permission by:

Curran Associates, Inc.
57 Morehouse Lane
Red Hook, NY 12571

CURRAN ASSOCIATES INC. proceedings.com

Some format issues inherent in the e-media version may also appear in this print version.

Printed by Curran Associates, Inc. (2017)

For permission requests, please contact the Institute of Physics
at the address below.

Institute of Physics
Dirac House, Temple Back
Bristol BS1 6BE UK

Phone: 44 1 17 929 7481
Fax: 44 1 17 920 0979

techtracking@iop.org

Additional copies of this publication are available from:

Curran Associates, Inc.
57 Morehouse Lane
Red Hook, NY 12571 USA
Phone: 845-758-0400
Fax: 845-758-2633
Email: curran@proceedings.com
Web: www.proceedings.com

Table of contents

Volume 825

14th High-Tech Plasma Processes Conference (HTPP 14)

3 July 2016, München, Germany

Accepted papers received: 14 March 2017
Published online: 12 April 2017

Preface

The High-Tech Plasma Processes Conference (HTPP) is a bi-annual international conference based in Europe with topics encompassing the whole area of plasma processing science. This conference is open to all the international community in the world involved in plasma science and plasma technology. The aim of the conference is to bring different scientific communities together, facilitate the contacts between science, technology and industry and provide a platform for the exploration of both fundamental topics and new applications of plasmas.

For this edition of HTPP, as was the case for the last, we have achieved a well balanced participation from the communities of both thermal and non-thermal plasma researchers. 75 people from 17 countries attended the conference with the total number of contributions being 74, consisting of 19 invited talks and 55 poster contributions.

As a HTPP tradition a poster competition has been carried out during the conference. The winner of the poster competition was Fabrice Mavier from Université de Limoges, France with his paper “Pulsed arc plasma jet synchronized with drop-on-demand dispenser”

All the participants also ejoyed the social program including an “unconventional” tour of the city, the visit to the famous Hofbräuhaus and the dinner at the Blutenburg, a beautiful inner-city castle.

We have received papers corresponding to the contributions of HTPP-2014 that have been submitted for publication in this volume of Journal of Physics: Conference Series. Each submitted contribution has been peer reviewed and the Editors are very grateful to the referees for their careful support in improving the original manuscripts.

In total, 18 manuscripts have been accepted for publication covering a range of topics of plasma processing science from plasma fundamentals to process applications through to experiments, diagnostics and modelling. We deeply thank the authors for their enthusiastic and high-grade contributions and we are convinced that this volume of Journal of Physics: Conference Series will be interesting for our community.

Finally, we would like to thank the conference chairmen, the members of the steering committee, the international scientific committee, the local organizing committee, the organizing secretariat and the financial support from the conference sponsors that allowed the success of HTPP-2016.

The Editors of the HTPP-2016 Proceedings

Dr. Dirk Uhrlandt, head of the ISC

Prof. Philippe Teulet

Prof. Jochen Schein

Neubiberg, 6th of March 2017

Conference photograph

HTTP14 was supported by

soyer

THE LINDE GROUP
Linde

ewm

ITIS

IOP Conf. Series: Journal of Physics: Conf. Series **825** (2017) 011002 doi:10.1088/1742-6596/825/1/011002

Peer review statement

All papers published in this volume of *Journal of Physics: Conference Series* have been peer reviewed through processes administered by the proceedings Editors. Reviews were conducted by expert referees to the professional and scientific standards expected of a proceedings journal published by IOP Publishing.

Arc tracking energy balance for copper and aluminum aeronautic cables

T André[1], F Valensi[1], P Teulet[1], Y Cressault[1], T Zink[2] and R Caussé[2]

[1] Université de Toulouse, UPS, INPT, CNRS, LAPLACE 118 route de Narbonne, F-31062 Toulouse cedex 9, France
[2] Airbus Operations S.A.S., Site de Saint Martin du Touch, 316 route de Bayonne, F-31060 Toulouse Cedex 9, France

Abstract. Arc tracking tests have been carried out between two voluntarily damaged aeronautic cables. Copper or aluminum conductors have been exposed to short circuits under alternating current. Various data have been recorded (arc voltage and current, radiated power and ablated mass), enabling to determine a power balance, in which every contribution is estimated. The total power is mainly transferred to the cables (between 50 and 65%, depending on the current and the cable type), and causes the melting and partial vaporization of the metallic core and insulating material, or is conducted or radiated. The other part is deposited into the arc column, being either radiated, convected or conducted.

1. Introduction

Fault arcs occurring in aeronautic electrical circuits may lead to significant electrical harnesses damages and should be eliminated or at least controlled to limit their consequences. In particular, short circuits occurring between two adjacent cables may lead to a particular phenomenon called "arc tracking" [1,2] corresponding to the propagation of the arc along the cables due to the cable metallic core ablation and to the surrounding dielectric material degradation. In traditional aircrafts, one solution consists in using copper cables with appropriate insulating material mainly composed of polytetrafluoroethylene (PTFE), a fluorocarbon polymer. New airplane concept introduces two strong constraints: lowering of weight and increase of the electrical power. The consequences of these constraints are the partial or total replacement of copper with aluminum in the cables, the increase of alternating current (AC), the strong increase of direct current (DC) parts and DC buses in the electrical circuit, and the replacement of metal by composite materials in the structure. These changes require the problem of electric arcs and arc tracking to be carefully considered again.

The present study aims to quantify the power transfers from the tracking arc to the environment in AC circuit at atmospheric pressure. It should be noted that there are very few academic works dedicated to the study of arc tracking. This lack of published data concerning the arc tracking phenomenon is associated to the difficulty to characterize the arc due its very unstable behavior. In the present wok, arc tracking tests are carried out between two aeronautic cables (each one connected to a phase of the AC circuit) following a specific procedure, and the main objective is to evaluate the various terms of power, based on several experimental measurements (current, voltage, radiated flux and mass loss) and on some theoretical analysis. The total power is calculated, as well as the power transferred to the cables (by means of the electrode voltage fall). The part causing the melting and vaporization of the cables is estimated, considering a total melting and a partial vaporization of the ablated matter, and then the power lost by conduction and radiation of the cables is inferred. Another part of the initial input power is transferred to the arc column. The radiated power of the plasma is then estimated, considering that the Vacuum Ultra Violet is absorbed in the very first microns of air, and the power lost by convection and conduction in the plasma is inferred.

IOP Conf. Series: Journal of Physics: Conf. Series **825** (2017) 012001 doi:10.1088/1742-6596/825/1/012001

2. Experimental setup

2.1. Material

The setup presented in Figure 1 was developed to create a fault arc between two cables for an adjustable period of time (from 10 ms to 10 s). The AC power supply, manufactured by Puissance Plus, has been specially developed for this study. It can be used in single-phase or two-phase (with 120 degrees phase shift) configuration and current is adjustable up to 350 A rms. The phase-to-neutral voltage can also be set, the maximum value being 230 V rms (400 V phase-to-phase). Then the frequency can be chosen between 380 and 800 Hz with a 2 Hz resolution.

The characteristics of the tested cables are given in Table 1. The first one is mainly made of aluminum (Cable 1) while the second one is mainly made of copper (Cable 2). Due to copper higher conductivity a smaller diameter for the later has been chosen (10 instead of 8 in the American Wire Gauge system) in order to get the same nominal current.

Type	AWG	Main material	Nominal current	Insulating layer thickness (mm)	Diameter of the metallic core	Mass proportion of metal
Cable 1	8	Aluminum	35 A	0.25	3.26 mm	81 %
Cable 2	10	Copper	35 A	0.25	2.59 mm	91 %
Cable 3	10		35 A	0.62	2.59 mm	75 %

Table 1. Characteristics of the various tested cable types.

For both cables the insulating material (polyimide layer wrapped with a sheath of PTFE) has a thickness of 0.25 mm. In order to study the effect of this parameter a third cable, similar to Cable 2 but with a 0.62 mm insulating layer thickness has been tested. All tests were performed with the association of two similar cable samples.

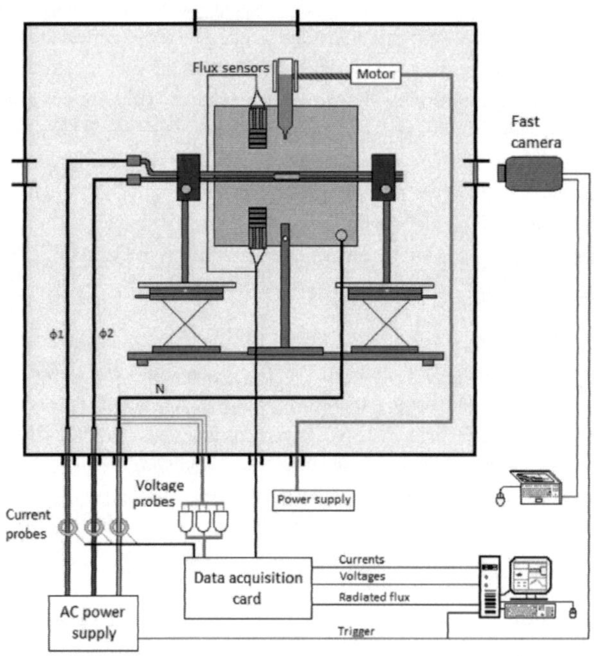

Figure 1. Experimental setup.

A specific procedure has been developed to cause arc ignition. For each experiment, the insulating sheath wrapping the metal core is partially stripped (one half of the cable circumference, for a length of 1 to 2 cm) on each cables. Then the two cables are bond together so that the two stripped areas are close to contact. Particular attention must be paid to this: if the contact between the two bare metal parts is too good there is just a short circuit, but no arc and if the spacing is too large ignition is not possible. The optimal distance is close to 1 mm. In the case of Cable 3 the sum of the thickness of the two insulating layers is about 1.2 mm and the distance is adjusted through deformation of the metal core (which is not solid, but constituted of small wires bundles). The insulating adhesive tape that bonds the cables is placed close to the striped area to avoid separation of the cables during the arc lifetime, due to electromagnetic force. The very erratic behavior of the arc can indeed lead to early arc extinction that is unwanted for our study and a tight cable holding is a good way to avoid this.

Each cable is connected to one phase. The ignition is facilitated with droplets of salted water (0.7% of NaCl) falling from a vial held above the stripped zone of the cables (this system works with a motor). The tests are carried out in a cylindrical chamber, at atmospheric pressure.

Tests in AC with a 230 V open-circuit phase-to-neutral voltage have been done for the three cable types (see Table 1) and for three current values: 175 A, 245 A, and 350 A. The frequency is 800 Hz. An aluminum plate (10cm × 10cm, 1.2 mm thick), representing the fuselage, is placed near the cables and connected to the neutral.

2.2. Measurements

Several kinds of measurements are done simultaneously: electrical measurements, radiation heat flux measurements and high speed imaging. The starting signal from the computer is used as a synchronization trigger for the power supply, the acquisition system (a National Instrument PXI 6221 board and an SCB 68 interface box) and the fast camera.

Electrical data are recorded for every test. The voltage between the two cables (connected to two different phases) is measured using DP-25 differential probes manufactured by Pintek. The voltage between each phase and the neutral (connected to the aluminum plate) is also measured with two additional similar probes. The current is measured in each phase and in the neutral using LF 505-S probes manufactured by LEM. Two heat flux sensors are placed around the cable samples along two perpendicular directions. They are positioned at a radial distance of 9.5 cm in a plane perpendicular to the cables axis, at the position where insulating sheath is removed. These sensors, manufactured by Captec, consist in thin strips of thermocouples (about 0.1 mm thick) arranged in a 1 square centimeter array. They deliver a voltage proportional to the heat input and are calibrated by the manufacturer so that the output voltage is proportional to the incoming radiated power. Each sensor is provided with its calibration data with a $10^{-3}\mu V$ accuracy (e.g. 1 $W.m^{-2}$ corresponds to 0.485 μV). All radiation from Vacuum Ultra Violet (VUV) limit (0.2 μm) up to far infrared (12 μm) are integrated. The temporal response of these sensors is about 80 ms and while it is not possible to observe fast fluctuations the duration of the tests (500 to 1000 ms) is sufficient to reach a stable value corresponding to the average radiated power. Measurements with an arcing time lower than 100 ms are not taken into account and the plateau is generally reached in a few hundreds of milliseconds. The sensors have to be frequently replaced: they are damaged by strong arc radiation and impacts of molten metal from the cables and their performances drop after only a few experiments. High speed imaging is performed with a Phantom V9 camera positioned outside the chamber, perpendicularly to the direction of the arc propagation. The recording speed is 1000 frame/s. Finally, the cables are weighed (with an accuracy of 0.1 mg) before and after each test, in order to determine the mass loss.

Due to the high instability of the arc, extinction often happened before the end of the selected time. Thus, many similar experiments have to be performed to get usable results.

3. Power balance

The considered power balance is presented below in Figure 2.

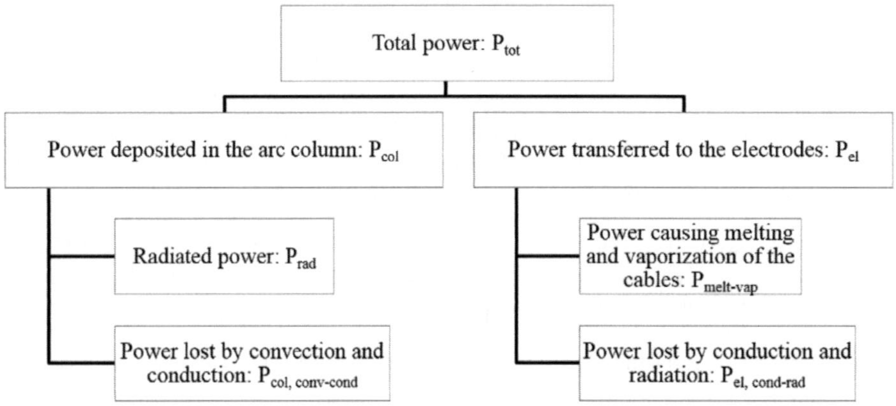

Figure 2. Power balance.

3.1. Total power

The electrical data recording allows calculating the total average power, as expressed in equation (1):

$$P_{tot} = \frac{1}{t_{arc}} \int u(t)i(t)dt \qquad (1)$$

where t_{arc} is the arc duration. The voltage u (between the two phases) can be expressed as in equation (2):

$$u = U_{el} + u_c \qquad (2),$$

where U_{el} is the electrode voltage fall (detailed below), and u_c is the arc column voltage. U_{el} is assumed to be not dependent on the current.

Part of this power P_{tot} is transferred to the cables constituting the electrodes, while the remaining part is deposited in the arc column.

3.2. Electrodes

The arc column voltage can be expressed as follows in equation (3):

$$u_c = E_c l \qquad (3),$$

where E_c is the average electric field within the plasma column, and l is the length of the arc column. The combination of (2) and (3) yields to the following equation (4):

$$u = U_{el} + E_c l \qquad (4)$$

We assume that E_c does depend neither on the current or the shape of the arc. Thus, the total voltage u is supposed to be a linear function of l. The extrapolation to a length equal to zero is used to get an estimation of the electrode voltage fall U_{el}. It yields to 18 V for copper cables, and 19 V for aluminum cables in the case of DC current as shown in figure 3. The accuracy of these values is rather low but they correspond quite well to values found in literature [2,3]. Similar measurements were performed in AC for the three tested currents, but the method is less accurate due to the more unstable behavior of the arc, the orientation of the plasma plume changing every half period. Values between 15 and 25 V were obtained and considering the error margins it has been decided to use the values obtained in DC current, since no more accurate data were found in literature concerning voltage fall in AC current.

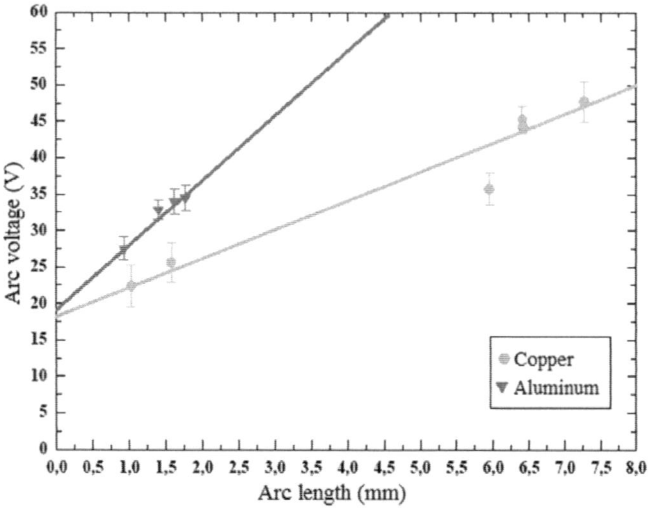

Figure 3. Determination of electrode voltage fall for copper and aluminum cables (DC current: 100 A). The solid lines correspond to liner fit of data.

We can then calculate the total average power P_{el} transferred to the electrodes, by multiplying the electrode voltage fall by the root mean square value of the current I_{RMS} (equation (5):

$$P_{el} = U_{el}.I_{RMS} \qquad (5)$$

Part of this power causes metal melting and vaporization, and can be expressed as in equation (6):

$$E_{metal} = E_{melt} + E_{vap} \qquad (6),$$

where E_{melt} is the energy required for the melting of the metal, and E_{vap} is the energy required for its vaporization. It is clear that all the ablated metal is melted but only a small part is vaporized. These two energies can be expressed as in the following equations (7) and (8):

$$E_{melt} = m.L_{melt} + \int_{T_{sol}}^{T_{melt}} m.C_{sol}(T).dT \qquad (7)$$

$$E_{vap} = m.L_{vap} + \int_{T_{melt}}^{T_{vap}} m.C_{liq}(T).dT = m.L_{vap} + m.C_{liq}(T_{vap} - T_{melt}) \qquad (8),$$

where T_{sol}, T_{melt}, and T_{vap} are the solid initial, melting and vaporizing temperatures, respectively; L_{melt} and L_{vap} are the latent heats of melting and vaporization respectively, C_{sol} and C_{liq} are the heat capacities at solid and liquid state, respectively, and m is the ablated metal mass (about 1 g for a 500 ms test at 244 A). C_{sol} varies with temperature, but C_{liq} remains constant [4]. Table 2 summarizes the thermodynamic data used to calculate the energy required for melting and vaporizing copper and aluminum. For better clarity, the energy per mass unit (in $J.g^{-1}$) to heat the solid material from ambient temperature (300 K) to melting temperature is given instead of solid heat capacity (in $J.g^{-1}.K^{-1}$). Only a small part of the ejected metal is vaporized, but the exact proportion is

IOP Conf. Series: Journal of Physics: Conf. Series **825** (2017) 012001 doi:10.1088/1742-6596/825/1/012001

unknown. Considering an analogy with plasma processes involving transient electric arcs and partial vaporization of electrodes (such as low voltage circuit breaker or lightning impact [5]), we assume that the amount of molten matter actually vaporized is 1%.

The insulating material ablation should also be taken into account. Since it is mostly composed of PTFE, we choose the value of 2.6 kJ/g for the heating, melting and evaporation of the ablated cable mass corresponding to the insulating material [6].

	Copper	Aluminum
Melting point (K)	1358	933
Boiling point (K)	2843	2791
Melting latent heat (J/g)	205	400
Boiling latent heat (J/g)	4725	10875
Energy for heating solid (J/g)	467	667
Liquid heat capacity ($J.g^{-1}.K^{-1}$)	0.517	1.177
Energy required for melting (kJ/g)	0.7	1.1
Total energy required for melting and vaporization (kJ/g)	6.2	14.1

Table 2. Thermodynamic data of copper and aluminum.

For those thermodynamic calculations, the mass proportion of metal and insulating material are taken into account. The power $P_{el,cond-rad}$, which is transferred to the cables but not involved in melting and vaporization, is lost by conduction and radiation. It is calculated as follows in equation (9):

$$P_{el,cond-rad} = P_{el} - P_{melt-vap} \qquad (9),$$

where $P_{melt-vap}$ is the power causing melting and partial vaporization of the cables.

3.3. Arc column

The total power P_{col} deposited in the arc column is calculated by subtracting the power transferred to the cables from the total power. Part of this power is radiated. The power flux measured by the flux sensors yields to the total radiated power assuming that the arc is punctual and by integrating the power density (assuming that it derives only from the arc column) over the surface of the sphere of radius equal to the average distance from the arc to the sensor (here 9.5 cm). The assumption of a punctual source is rather good considering the size of the plasma (around 1-3 mm). The obtained signal is converted according to the sensors calibration so that we get directly the radiated power as a function of time. The average radiated power P_{rad} can then be expressed as follows in equation (10):

$$P_{rad} = 4\pi d^2 P_{meas} \qquad (10)$$

where d is the distance between the arc and the sensors, and P_{meas} is the average value given by the flux sensors. It can be either the average value of the two sensors if the responses are close to one another, or only the one which gives the highest values (one of the two sensors could be partially hidden by a cable), over the period where the value of measured radiated power reaches a plateau.

Since the experiments have been performed in air at atmospheric pressure, all the radiation with a wavelength lower than 200 nm (called Vacuum Ultra Violet or VUV radiation) is absorbed in the ambient air within a few microns, mainly because of photodissociation and photoionization of oxygen molecules. Due to these phenomena, the radiation measured by our sensors (located a few cm from the arc) does not include the VUV part of the radiation directly emitted by the arc. Since this part is far from being negligible (it could be equal or even greatly higher than non-VUV radiation) it is mandatory to get information on this contribution. Total

IOP Conf. Series: Journal of Physics: Conf. Series **825** (2017) 012001 doi:10.1088/1742-6596/825/1/012001

removal of oxygen is difficult to achieve (even by filling the chamber with nitrogen, the gas will still contain some oxygen as impurities). Moreover, even for the non-VUV part the uncertainties associated to radiation flux measurement (see figure 6) are quite high. Then the calculation of the Net Emission Coefficient (NEC) [7,8] appears to be a good complementary approach to determine the total radiated power that allows getting VUV contribution and comparing non-VUV contribution to measurements. The NEC is computed as the divergence of the total radiation flux at the center of an isothermal sphere of radius R_p (power per volume unit and per solid angle unit). Radiation power is integrated over a large range of photon energy corresponding to the wavelength range from 30 nm to 5 μm. Two examples (99% air + 1% Cu or 1% Al) are presented in Figure 4, showing the variation of the total NEC and the partial NEC (i.e. for λ>200 nm, which does not include the VUV part) for a plasma radius R_p =2mm as a function of temperature. In our case one can assume that the mean temperature in the core of the plasma is close to 15 000 K, by comparison to thermal plasmas in similar conditions such as encountered in low voltage circuit breakers [9]. The results show that for both materials the total NEC is similar (~3.0^9 $W.m^{-3}.sr^{-1}$ at 15000 K), as seen in figure 4. However the contribution of the non-VUV part is not the same: while it is 30 % for copper it is only 15 % in the case of aluminum. However this strong predominance of VUV radiation has little impact on our energy balance: even if the VUV radiation is not directly measured by the flux sensors, it is indirectly taken into account in the other power terms. Indeed, the power absorbed by oxygen molecules (VUV radiation) is either emitted again in the visible light or in the infrared radiation range (so it is detected by the flux sensors), or lost by convection and conduction (heating of the gas and of the plasma, which is also considered in our balance). All our determination of radiated energy (by mean of heat flux sensors or by calculation) have to be considered carefully since they present quite large uncertainties (due to the simplifying assumptions considered for calculation and to sensor noise level for measurements, see figure 6).

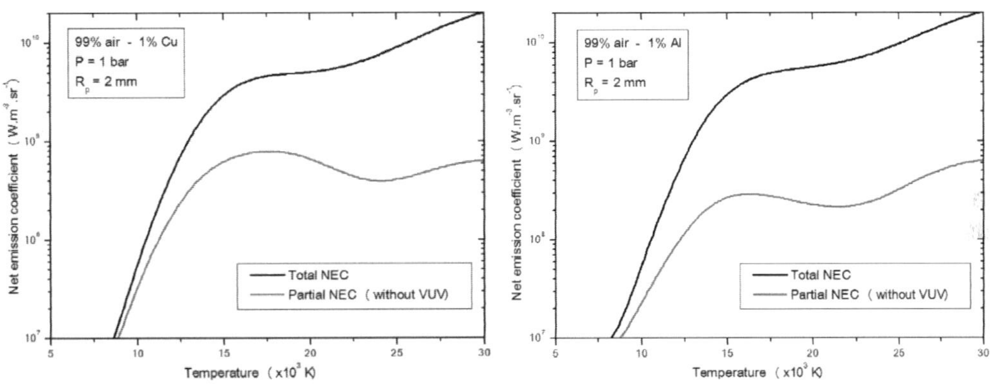

a) b)

Figure 4. Comparison between the NEC and the partial NEC (without VUV) for mixtures of 99% air and a) 1% Cu, b) 1% Al (mass proportions), and for a 2mm R_p (plasma radius).

Another part of the power transferred to the arc column is lost by convection and conduction. This power, $P_{col,conv-cond}$ dissipated in the arc column is inferred by subtraction as follows in equation (11):

$$P_{col,conv-cond} = P_{col} - P_{rad} \qquad (11)$$

4. Results and discussion

Table 3 presents a synthesis of the power balance for the considered cables and currents. The results for each case correspond to an average of several tests (at least 10) realized in these conditions. The uncertainties given for erosion rate (in g/s), erosion speed (in g/cm, calculated from the erosion rate by considering the linear mass) and the associated power $P_{melt\text{-}vap}$ arise from statistical dispersion of the results on mass loss measurement. The uncertainties concerning total power P_{tot} and the power at the electrodes P_{el} (10 to 15%) result from the uncertainties associated to voltage fall measurement. The uncertainties for $P_{el,cond\text{-}rad}$, P_{col} and $P_{col,conv\text{-}cond}$ do not appear in Table 3, since these values are not experimental data but the results of subtraction on the other terms. Concerning the radiated power P_{rad}, the value of 1300 W has been used for all cases. This choice is based on power flux measurement and calculation (see below). Both methods lead to quite high uncertainty but the combination of the two allows estimating it to about 20% and since the difference between copper and aluminum is low it was not relevant to consider it. Most of the total power (P_{tot}) is transferred to the electrodes (part corresponding to P_{el}) but this proportion decreases with current, from 65% at 174 A to 53% at 350 A in the case of cable 1. It is larger for copper than for aluminum (65% and 56% for cable 1 and 2, respectively) at 244 A but there is almost no difference at higher current. Since they are based on the electrical measurements, which present good accuracy and time resolution, the total average power and the power transferred to the electrodes seem to be the most reliable calculated results. The erosion speed increases with current and is greater for copper than for aluminum. However the use of a thicker insulting layers (cable 3) notably reduces it.

Cable	**I rms (A)**	**Erosion rate (g/s)**	**Erosion speed (cm/s)**	P_{tot} **(W)**	P_{el} **(W)**	$P_{melt-vap}$ **(W)**	$P_{el,cond-rad}$ **(W)**	P_{col} **(W)**	P_{rad} **(W)**	$P_{col,conv-cond}$ **(W)**
Cable 1	174	0,9±0.3	2.31± 0.8	5099±750	3306±330	1318±439	1988	1793	1300	493
Cable 1		1,5±0.2	3.85± 0.5	7630±900	4636±460	2196±293	2440	2994	1300	1694
Cable 2	244	2,1±0.5	4.20± 1	6738±1000	4392±440	1881±447	2511	2346	1300	1046
Cable 3		2,1±0.2	3.39± 0.3	7898±400	4392±440	2510±239	1882	3506	1300	2206
Cable 1		2,3±0.3	5.90± 0.8	12502±1100	6650±670	3367±439	3283	5852	1300	4552
Cable 2	350	3,2±0.3	6.40± 0.6	11501±1000	6300±630	2866±269	3434	5201	1300	3901
Cable 3		3,3±0.3	5.32± 0.5	12246±1100	6300±630	3944±356	2356	5946	1300	4646

Table 3. Power balance

The electrical measurements give information not only on voltage and intensity, but also on the occurrence of discharge and on the potential transfer of the arc to the aluminum plate. In Figure 5, a typical case is presented. In a first step (from 0.25 up to 0.4 s) the arc occurs between the cables, so there is no current in the neutral, and the voltage between each phase and the neutral is equal to the open circuit phase-to-neutral voltage (i.e. around 325 V peak value). In a second step (from 0.4 up to 0.55 s), the arc transfers to the aluminum plate, so the current in the neutral suddenly grows to a non-zero value, and the voltage between each phase and the neutral drops during the transfer time. Then the transfer stops (from 0.55 up to 0.75 s) and the arc occurs again only between the cables. The total duration of the test is 0.5 second.

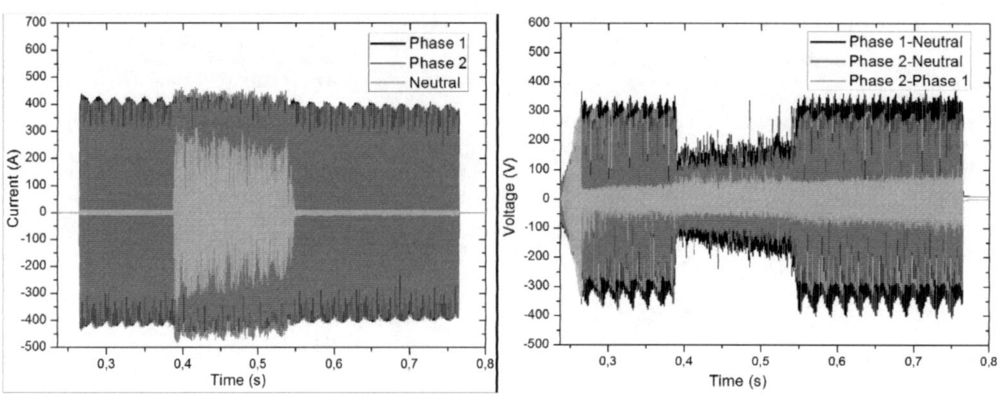

Figure 5. Electrical data from a test on an aluminum cable with a 350A current input. The power transfer of the arc to the aluminum plate can be noticed between around 0.4 and 0.55 s.

About half of the power transferred to the cables corresponds to the power needed to cause their melting and partial vaporization. This proportion increases with current (from 40% at 174 A to 51 % at 350 A in the case of aluminum) and seems to be higher for aluminum than for copper (47% and 43%, respectively, at 244 A). This result is obtained assuming that the same proportion of metal (1%) is vaporized for both cable. However the larger fume production in the case of copper cables might be related to higher metal vaporization rate, which means that the proportion is underestimated in the case of copper. In the case of a large insulating layer (cable 3) the part corresponding to cable ablation becomes dominant, especially at high current. We observe very often that molten metal droplets remain on the cables and solidify. These droplets are not taken into account in the ablated mass measurement, so the power causing the melting of the corresponding amount of metal is missing in the balance.

The data given by the flux sensors do not exhibit a real influence of the current and the type of cable on the radiated power. As an example the radiated flux for cable 1 at 174 A and for cable 2 at 244 A are given in figure 6. In both cases the flux is measured at about 10500 W/m², which corresponds to a power of 1200 W. The horizontal light grey line represents the considered value, corresponding to the average between sensors 1 and 2 at the plateau indicating stable configuration.

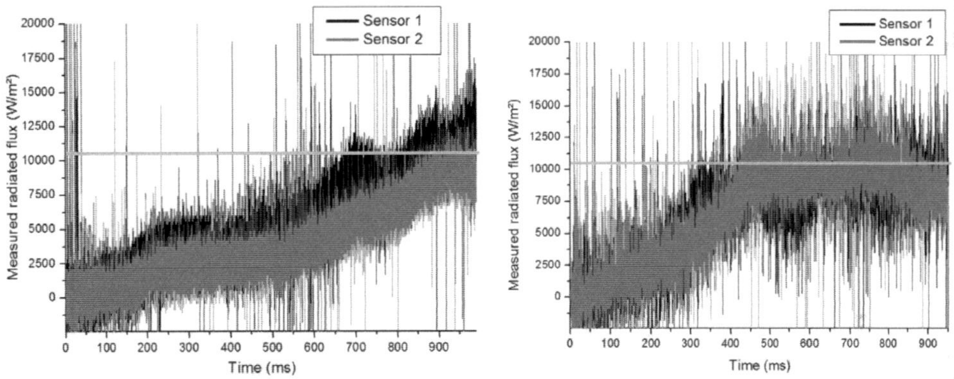

Figure 6. Measured radiated flux: a) cable 1 (aluminum) at 174 A; b) cable 2 (copper) at 244 A.

These results were compared to the estimation provided by the NEC. Considering that the plasma radius R_p is equal to 2 mm, the height of the arc h = 3 mm, that the mean temperature of the plasma core is approximately 15000 K, and that the plasma consists in 99% air and 1% Cu or 1% Al (thus the NEC ε_N is around 3.10^9 $W.m^{-3}sr^{-1}$), we obtain:

$$P_{rad,calc} = 4 \cdot \pi^2 \cdot {R_p}^2 \cdot h \cdot \varepsilon_N \approx 1400\ W \qquad (12)$$

Then even if the signal from heat flux sensors is rather noisy and erratic both results are compatible and the average value of 1300 W used in Table 3 seems to be a reasonable approximation. The proportion of the power deposited in the arc column corresponding to radiated power decreases with current and is larger for copper than for aluminum. The use of thicker insulating sheath strongly decreases the ratio at 244 A. At high current the differences between the three cables reduce and the ratio does not exceed 25%, while at 174 A it is greater than 70% for aluminum. For the tests involving cables 1 samples, we can observe a white sheath at the surface of the projected metal droplets, which could be alumina Al_2O_3. As the reaction of oxidation of aluminum is highly exothermic, this phenomenon could modify the power balance, but it is difficult to estimate this additional input of energy.

5. Conclusion and perspectives

Short-circuit tests of arc tracking have been performed between two aeronautic cable samples, in copper or aluminum under alternating current and at atmospheric conditions. A power balance has been realized, which shows that about half of the total power is transferred to the cables, causing their degradation (melting and partial vaporization of the metallic core and degradation of the insulating material). This proportion is greater for aluminum than for copper but in the latter case it might be underestimated, since stronger vaporization seems to occur in the case of copper cables. Besides, since the erosion speed is lower, for the same input energy the length of damaged cable will be smaller in the case of aluminum. One can also note that the use of a thicker insulating layer helps reducing the damages, since for similar energy transferred to the electrodes for cable 2 and cable 3 the erosion speed is lower for cable 3. In addition, the fact that a larger part of energy is associated to cable degradation (greater quantity of insulating material) means that there is less energy available to cause other damages, such as on the surrounding structure. This power deposited in the electrodes can be divided into two parts, the main one corresponding to the melting and vaporization of the cables and the other being lost in material (heat conduction losses in the cables and thermal radiation). Another part of total input power is deposited in the arc column, where it is either radiated, or lost by convection or conduction. The power emitted in the VUV range is not received by the flux sensors, as it is immediately absorbed, but it is taken into account in the balance, because it is either reemitted (at greater wavelength: near IR mainly) or involved in plasma heating. A next step will consist in performing the same tests under reduced pressure (0.1 bar), to represent an aeronautic configuration in a better way (cruise altitude) and investigate the behavior of the arc in such conditions. Moreover, the current energy balance does not take into account the effect of any chemical reaction. In particular the oxidation of aluminum is a very exothermic reaction leading to possibly non negligible additional energy input. Further work will be dedicated to the study of formation of alumina and the influence of the oxidizing nature of the plasma forming gas (tests in pure nitrogen).

References

[1] Dricot F and Reher H J 1994, Survey of arc tracking on aerospace cables and wires, *IEEE Transactions on Dielectrics and Electrical Insulation* **1** pp 896–903.

[2] El Bayda H, Valensi F, Masquère M and Gleizes A 2013, Energy losses from an arc tracking in aeronautic cables in DC circuits, *IEEE Transactions on Dielectrics and Electrical Insulation* **20** pp 19-27

[3] Agarwal M S and Holmes R 1984, Arcing voltage of the metal vapour vacuum arc, *J. Phys. D: Appl. Phys.* **17** pp 757-767

[4] Chase M W 1998, NIST-JANAF Thermochemical Tables. Washington, DC: New York: American Chemical Society ; American Institute of Physics for the National Institute of Standards and Technology.

[5] Lago F, Gonzalez J J, Freton P and Gleizes A 2004, A numerical modelling of an electric arc and its interaction with the anode: Part I. The two-dimensional model, *J. Phys. D: Appl. Phys.* **37** pp 883–897

[6] Godin D, Trépanier J Y, Reggio M, Zhang X D and Camarero R 2000, Modelling and simulation of nozzle ablation in high-voltage circuit-breakers, *J. Phys. D: Appl. Phys.* **33** pp 2583-2590

[7] Naghizadeh-Kashani Y, Cressault Y and Gleizes A 2002, Net emission coefficient of air thermal plasmas, *J. Phys. D: Appl. Phys.* **35** pp 2925–2934

[8] Lowke J J 1974, *J. Quant. Spectrosc. Radiat. Transfer* **14** p 111

[9] Swierczynski B, Gonzalez J J, Teulet P, Freton P and Gleizes A 2004 Advances in low-voltage circuit breaker modelling, *J. Phys. D: Appl. Phys.* **37** pp 595–609

Simulation of the Particle Melting Degree in Air Plasma Spraying

K. Bobzin[1], M. Öte[1], M. A. Knoch[1], I. Alkhasli[1]*, U. Reisgen[2], O. Mokrov[2], O. Lisnyi[2]
[1] RWTH Aachen University, IOT - Surface Engineering Institute, Aachen, Germany

[2] RWTH Aachen University, ISF – Welding and Joining Institute, Aachen, Germany

E-mail: alkhasli@iot.rwth-aachen.de, *corresponding Author

Abstract. Plasma spraying is a coating process which is widely used for the application of thermal barrier coatings. High plasma jet temperatures allow the processing of ceramic material particles which characteristically exhibit low thermal conductivities. This, in turn, produces high temperature gradients inside the particles and vaporization on the particles' surface during their dwell time in the plasma jet. Thus, a single particle in the plasma-jet can exhibit 3 states of matter simultaneously: solid in the core, molten exterior and gaseous on the surface. The temperature distribution inside the particles is the foremost factor which influences the particles' behavior during their impact on the substrate surface. Experimental investigations can provide only the surface temperature of the particles, which is not a good indicator of the melting degree for ceramic particles due to the high temperature gradients. This study focuses on the determination of the temperature distributions inside the particles, during their flight in the plasma and the free jet, with the help of numerical simulations. For this purpose, a numerical model of the plasma spraying process that describes the plasma and the free jet loaded with sprayed particles is presented. The model includes three sub-models; a plasma torch sub-model, a particle-laden free jet sub-model and a powder particles sub-model. The first sub-model calculates the temperature and velocity fields of the plasma inside the plasma torch, the second sub-model the kinetics of temperature and velocity fields of the turbulent free jet generated by the plasma torch. This information is used in the third sub-model to calculate the kinetics of temperature distribution within the particles, their melting degree and mass losses due to evaporation. The third sub-model also calculates heat and mechanical impulse loses due to the particle-plasma interaction, which in turn is coupled with the free jet sub-model.

1. Introduction

The air plasma spraying process (APS) has been the subject of numerous numerical models which attempt to consider the whole complexity of the relevant physical phenomena. The simulation of a single cathode plasma spraying systems involves transient models of the arc root fluctuations at the anode [1, 2] and the effect of plasma fluctuations on the particle temperature and velocity [3, 4]. In a three-cathode APS system, the arc root fluctuations are vastly diminished by a special placement of neutrode rings and the anode. This allows steady-state simulations of such systems [5, 6]. Steady-state simulations of a three-cathode APS system often consider the dispersion of the particles at the injector exit, the influence of the powder carrier gas flow on the plasma flow as well as the particle distribution in the plasma jet [7]. Relevant are the distributions of the size, the velocity and the temperature of the

IOP Conf. Series: Journal of Physics: Conf. Series **825** (2017) 012002 doi:10.1088/1742-6596/825/1/012002

particles in the plasma jet [6, 8]. The most common approach for the determination of the particle temperatures is the lumped capacitance method which assumes a homogenous temperature within the particle [9, 10, 11]. For materials with a high thermal conductance such as metals, especially for particles smaller than 100 µm in diameter, this assumption seems plausible and delivers results close to reality. However, ceramic materials, which are commonly used in APS applications, usually have a significantly lower thermal conductance than metals [12]. This causes substantial temperature gradients between the particle surface and its core. A ceramic particle which is molten on its surface but still has a solid core is a commonplace phenomenon in APS which was taken into account in various simulations by considering heat transfer within the particles [13, 14, 15, 16, 17, 18, 19]. An inhomogeneous melting of ceramic particles has a significant effect on the bonding behaviour and the porosity of the final coating and is therefore considered in this study.

State of the art particle diagnostics in thermal spraying does not provide information regarding the temperature gradients in the particle [20]. Temperature measurements, which are based on the two-colour pyrometry, rely on the emission values on the particle surface [21]. Accordingly, they deliver only the surface temperature of the particle. It's a known case, that an increase of the electrical current in the plasma generator might deteriorate the deposition efficiency of the spraying process. The reason is that increasing the current increases not only the plasma temperature, but also the plasma velocity. This, in turn contributes to the increase in particle velocities and therefore, to the decrease of the particle dwell times in the plasma jet [6]. In such cases, experimental measurement of the particle temperature indicates temperatures above the melting point of the spray material, nonetheless the solid core springs back off the substrate surface. The aim of this study is to develop a numerical model for the heat conduction and the phase change phenomena in order to determine the temperature distribution and the location of the melt front within the particles during their flight in the plasma jet. This will contribute to the understanding of the bonding behaviour of ceramic particles in APS which in turn would enable the prediction of deposition efficiency and coating porosity. In the next section, the modelling of the plasma and the free jet will be described, followed by the description of the modelling of sprayed particle. Input parameters that are used in the simulations are presented in the "Model Parameters and Boundary Conditions" section. The results of the simulations and their analysis are presented in the "Results and Discussions" section which is followed by the Conclusion.

2. Modeling of Plasma and Free Jet

In this section, the numerical model that was used to describe the physical phenomenon in the plasma torch and the free jet as well as the boundary conditions will be described. The underlying principle of a plasma torch is based on the ionisation of a working gas between the electrodes by applying an electrical potential difference between the electrodes. This causes the gas to ionise and form an electrically conductive path between the electrodes. Thus, the conversion of the electrical energy into heat energy is achieved in the plasma generator. After being converted into the plasma state, the process gas leaves the nozzle of the plasma generator at high temperatures (T_P > 12,000 K) and velocities (v_P > 1,000 m/s) [7].

2.1 Boundary conditions and model parameters

In this work, the three-cathode plasma torch TriplexPro^{TM}-210 from Oerlikon Metco is analysed. It has 24 radially placed inlets which pass the operating gas into the convergent part of the plasma torch while the inclination of the inlets causes the gas to swirl. The anode ring which is located next to the torch outlet is separated from the cathodes with a long tubular section of neutrode rings. The geometry of the interior of the plasma generator is depicted in the Figure 1. The plasma torch allows the injection of the particles into the plasma flame from three radially placed injectors downstream to the outlet.

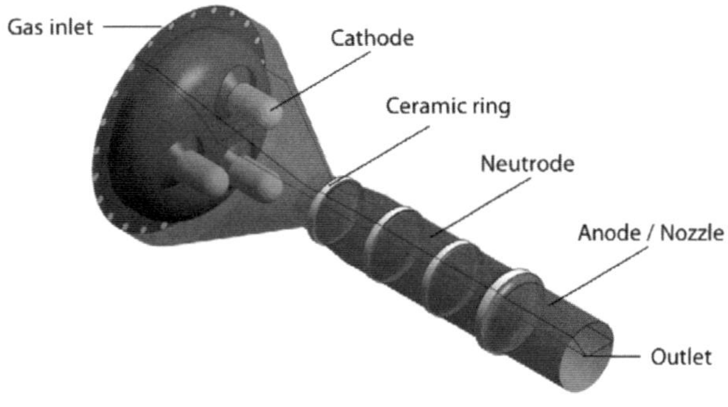

Figure 1. Interior geometry of the plasma generator TriplexPro™-210 torch from Oerlikon Metco GmbH (Kelsterbach, Germany)

The plasma simulation in the plasma torch is already an established model [5–7] and is coupled with the free jet simulation [8]. The boundary conditions of the free jet are the plasma flow profiles at the plasma torch outlet. The flow profiles are exported from the plasma torch simulation and imported into the free jet simulation as a boundary condition. The flow properties that are imported to the free jet simulation at the plasma torch nozzle are the velocity profile, total temperature, turbulent kinetic energy and turbulent eddy dissipation.

Generally speaking, there are several approaches to plasma modelling which include single particle, kinetic, fluid, hybrid kinetic/fluid, gyrokinetic and system of many particles. Kinetic and fluid models are the most widely used ones in the literature [2, 3, 5, 6]. While the kinetic description of the plasma delivers more accurate results, due to its complexity and the computational overhead associated with it, the fluid model of the plasma is considered in this work. The fluid model describes the plasma based on the macroscopic parameters, which are determined by coupling the mass, momentum and energy conservation equations with the Maxwell equations. A commercial general-purpose CFX package Ansys CFX® (Release V15) was employed for the calculation of the plasma behaviour in this study. The following system of equations has to be solved numerically in order to describe the plasma flow:

$$\frac{\partial \rho}{\partial t} + \nabla(\rho\vec{u}) = 0, \tag{1}$$

$$\rho\left(\frac{\partial \vec{u}}{\partial t} + \vec{u}\nabla\vec{u}\right) = \rho\vec{g} + \nabla\vec{\tau} - \nabla p + \vec{f_L} + S_V, \tag{2}$$

$$\frac{\partial \rho h}{\partial t} + \nabla(\rho h\vec{u}) = \nabla\left(\frac{\lambda}{C_p}\nabla h\right) + S_{Joule} + S_{Net} + S_H, \tag{3}$$

$$S_{Net} = S_{Rad} + S_{Abs} \tag{4}$$

The first three equations describe the thermodynamic fluid flow of the plasma and are equations of conservation of mass (Eq. 1), momentum (Eq. 2) and energy (Eq. 3) and are coupled with the Maxwell equations of electromagnetism. The coupling between the electromagnetic forces and the fluid flow dynamics is defined by the Lorentz force, the Joule heating which describes the electric heating of the process gas in the plasma torch as well as the radiation model. Radiation is modelled according to the temperature and pressure dependent net emission coefficient model. The values for the coefficient is taken from the literature [22, 23]. The momentum loss of the free jet due to the particle is taken into account with the term S_V. The term S_H in the Eq.(3) corresponds to the heat exchange between the plasma jet and the particle. S_V and S_H will be addressed in chapter 3.4 in more detail.

In order to predict the magneto hydrodynamic behaviour of the plasma column in the plasma generator, the temperature and pressure dependent properties of the plasma gas are included in the calculations. Properties of the most common operating gases in thermal spray are available in the literature [24, 25]. In order to model the turbulent mixing and the heat dissipation in the turbulent plasma flow, the two-equation turbulence SST model is used.

Because of the high plasma temperatures, the losses due to the radiation are significant in the total heat balance of the plasma flow. The net emission coefficient model was employed for the calculation of the radiation losses according to [5, 6]. The radiation energy balance is defined as the difference between the emitted and the absorbed radiation which is in turn a temperature and pressure dependent term of the energy conservation equation. This model calculates the radiation losses according to the temperature and pressure distribution in the simulation domain. The radiative warming of colder regions is neglected due to the low radiative absorption of the surrounding gases. The input parameters for the plasma jet model are given in the Table 1:

Table 1. Input parameters for the particle-laden plasma-jet model

Property	Unit	Value
Plasma gas		Argon
Arc current	A	500
Gas flow rate	SLPM	50
Nozzle diameter	mm	9
Carrier gas flow rate	g/s	0.15
Particle injection rate	g/s	0.4
Particle mean diameter	μm	38
Particle diameter standard deviation	μm	11

3. Modelling of Sprayed Particles

The mathematical model of the particle behaviour in the plasma jet is described by the equations of transient heat transfer, balance of mass, movement and interaction conditions with the free jet. The main assumptions that are used for the particle description are listed below:

- spherical particle geometry
- particle temperature distribution has a spherical symmetry
- local plasma temperature and velocity along the boundary layer of particles are constants and defined by the current particle mass center position in the free jet
- particle size can vary due to evaporation
- particle evaporation has no influence on heat exchange between the particle and flow jet
- temperature and velocity fields of the free jet are given and calculated according to a-priori simulations as shown in [5, 6].

Besides the boundary conditions, each particle is defined by a set of initial conditions. The most important ones are:

- initial radius R_0
- initial temperature T_0
- coordinates of the injection (point A in Figure 2)
- initial particle velocity at point A

Additional parameters are the volume flow rate of the carrier gas, diameter of the particle injector, particle mass feed rate, maximal distance of the particle movement along the free jet direction and the temperature dependent material properties of the particle (Table 3). The coefficients and constants used for the free jet and the particle heating simulations are listed in Table 2.

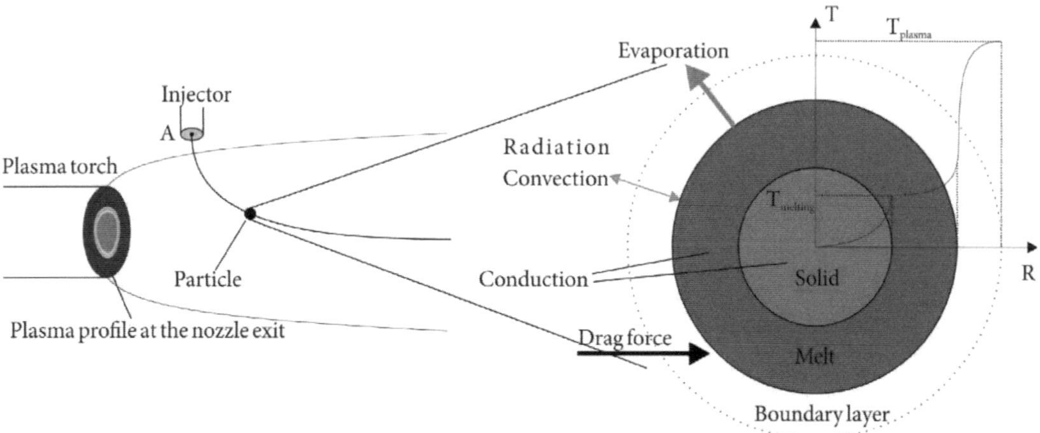

Figure 2. Schematic description of the discrete particle traveling in free jet and the particle heating model

Table 2. Coefficients and constants used in the simulations

Property	Symbol, unit	Value
Stefan-Boltzmann constant	σ, W cm^{-2} K^{-4}	5.67 10^{-4}
coefficient of radiation emission	ε_σ	0.6
gravitational acceleration	g, cm s^{-2}	980.0
ambient temperature	T_0, K	293.15
ambient pressure	p, bar	1.013

3.1 Particle heating

Particle heating in a free jet is described by the heat equation written in spherical coordinates, with its origin located at the particle center [26]. The heat equation is written one dimensional form due to the spherical symmetry assumption of the particles.

$$\frac{\partial H}{\partial t}=\frac{1}{r^2}\frac{\partial}{\partial r}\left(r^2\lambda\frac{\partial T}{\partial r}\right),\qquad(5)$$
$$0<r<R,\ \ t_A<t<t_L$$

$$H=\rho\int_{T_0}^{T}c\,dT+\rho\chi_m\qquad(6)$$

Here, $H(T)$ – is enthalpy, T – temperature, R – the instantaneous radius of the particle, t - the particle heating time, t_A –time at the point A (Figure 2), t_L – time at the axial distance L and λ - thermal conductivity. Enthalpy is described by the Eq. (6), where c – is heat capacity, ρ – density, χ_m - latent melting heat. Boundary conditions are defined as:

$$\left.\frac{\partial T}{\partial r}\right|_{r=0} = 0 \tag{7}$$

$$\lambda_p \left.\frac{\partial T}{\partial r}\right|_{r=R} = q - \chi_v g_v \tag{8}$$

$$q = \alpha(T_g - T(R)) + \xi\sigma_0(T_g^{\ 4} - T(R)^4) \tag{9}$$

where q is the heat flow from the gas into the particle, with the coefficient of convective heat exchange between the particle and free jet α and radiative heat exchange via the emissivity factor ξ. T_g is the gas temperature at the particle location and $T(R)$ is the particle surface temperature. χ_v is the heat of vaporization, $g_v(T)$ - specific surface mass flow rate of vaporization and σ_0 the Stefan-Boltzmann constant.

3.2 Particle size

Since the evaporation at the particle surface was taken into account, the particle radius may change. At each time increment the instantaneous particle radius defined by the mass balance equation:

$$\frac{dm}{dt} = g_v S \tag{10}$$

with the particle mass $m = 4/3\pi R^3\rho$ and the particle surface areas $S = 4\pi R^2$. The impact of the surface evaporation on the overall particle mass will be a subject of further studies.

3.3 Particle velocity and trajectory

The particle velocity and trajectory can be obtained by the equating Newton's second law to the aerodynamical drag force acting on the particle surface during its flight in the free jet. For this purpose, the velocity of discretely defined particles is determined as the time derivative of the particles' centre coordinates. The equations are written as follows:

$$\frac{d(m\vec{v})}{dt} = \vec{\mathbf{F}} \tag{11}$$

$$\vec{\mathbf{F}} = C_d S_m \frac{\rho_g (\vec{u} - \vec{v})|\vec{u} - \vec{v}|}{2} + m\vec{g} \tag{12}$$

where $\vec{v}$ – is the particle velocity; $\vec{u}$ – is the local free jet velocity which takes into account the interaction with sprayed particle; $\vec{\mathbf{F}}$ – is the aerodynamical drag force vector; C_d – is the drag coefficient of a spherical particle; $S_m = \pi R^2$ – is the cross-sectional area of a spherical particle; ρ_g - is the density of surrounding gas; $\vec{g}$ - is the gravitational acceleration.

Different approaches exist for the description of the coefficients of convective heat exchange and aerodynamical drag for a spherical particle. A comparison regarding the influence of the different descriptions of the coefficients is presented in [27]. In this work, the description of coefficients according to [26] is employed as depicted in equations (13-17).

$$\alpha = Nu\frac{\lambda_g}{2R} \tag{13}$$

$$C_d = \begin{cases} 24\,\mathrm{Re}^{-1}, & \mathrm{Re} \le 0.2 \\ 24\,\mathrm{Re}^{-1} + 3.6\,\mathrm{Re}^{-0.317}, & 0.2 < \mathrm{Re} \le 4.0 \\ 24\,\mathrm{Re}^{-1} + 4\,\mathrm{Re}^{-0.333}, & 4.0 < \mathrm{Re} \end{cases} \tag{14}$$

$$Nu = 2\frac{\lambda_g}{\lambda} + 0.5\,\mathrm{Re}^{0.5}\,\mathrm{Pr}^{0.4}\left(\frac{\rho_g\big|_{r>>R}\,\eta_g\big|_{r>>R}}{\rho_g\big|_{r=R}\,\eta_g\big|_{r=R}}\right)^{0.2} \tag{15}$$

$$\mathrm{Re} = 2R\,\rho_g\left|u - v\right| / \eta_g \tag{16}$$

$$\mathrm{Pr} = C_g \eta_g / \lambda_g \tag{17}$$

where Nu - is the Nusselt number; Re - the Reynolds number; Pr- the Prandtl number; η - dynamic viscosity of the free jet; λ_g, ρ_g, η_g, C_g - are the free jet properties at the point of the particle location, calculated at the particle surface temperature.

3.4 *Influence of the sprayed particles on the free jet*

The particles injected into the free jet are accelerated and heated by the gas. At the same time the free jet is slowed and cooled down due to the influence of the injected particles [28]. In order to take into account this effect the "two-way coupling" method is implemented. This is achieved by coupling the equations of momentum (equation (2)) and energy (equation (3)) of the fluid flow with the momentum and energy equations of the particle with the help of the terms S_V and S_H. These terms incorporate the momentum and energy loss due the acceleration and heating up the particles into the free jet flow and are defined as:

$$S_v = \frac{1}{\delta V}\sum_{k=1}^{b} \vec{\mathbf{F}}_k S_k \tag{18}$$

$$S_H = \frac{1}{\delta V}\sum_{k=1}^{b} q_k S_k \tag{19}$$

$$n_t = \frac{3}{4\pi R_0^3}\frac{G_M}{\rho} \tag{20}$$

where S_V, S_H are the local values of the free jet heat and mechanical impulse loss due to the interaction of the free jet with the particles; b – is the instantaneous number of particles that are locally present in a free jet volume δV; n_t – is the integer part of the amount of the particles homogenously distributed on the diameter of the injector nozzle and G_M is the particle feed rate. Initial condition of the particle loading of the plasma flow is defined by $S_V = 0$, $S_H = 0$, $G_M = 0$.

3.5 *Material properties used in the particle heating model*

The model described by equations (5 - 20) enables the calculation of the particle velocity, particle trajectory and kinetics of radial temperature distribution in the particle, the ratio of its melting and mass losses due to evaporation. These particle parameters can be calculated in dependence on the material composition, initial temperature and the particle diameter, particle feed rate and further conditions of particle injection into the plasma jet. Besides, this model can be used to observe the behaviour of multiple particles simultaneously. The material properties of the powder particle used in the simulation were chosen according to [28] and are shown in Table 3:

Table 3. Material properties of the feedstock material [28]

Property	Symbol, unit	Value
Material		Al_2O_3
Melting temperature	T_M, K	2,345
Boiling temperature	T_B, K	3,250
Latent heat of melting and solidification	χ_m, J g^{-1}	1,150
Latent heat of vaporization	χ_v, J g^{-1-}	19,380
Material density	ρ, g cm^{-3}	3,690
Heat capacity	c , J g^{-1} K^{-1}	1.2
Thermal conductivity	λ, W cm^{-1} K^{-1}	0.3

4. Results and Discussion

A plasma generator simulation was conducted using the parameters indicated in the section 2.1. The velocity and temperature fields at the plasma generator nozzle exit were transferred to the free jet simulation as boundary conditions. Flow conditions at the nozzle exit served as an inlet condition for the free jet computational domain. Figure 3 illustrates the trajectories of the particles injected into the free jet with an indication of the particle temperatures during their flight. Particle temperatures shown in the figure were obtained using the lumped capacitance method.

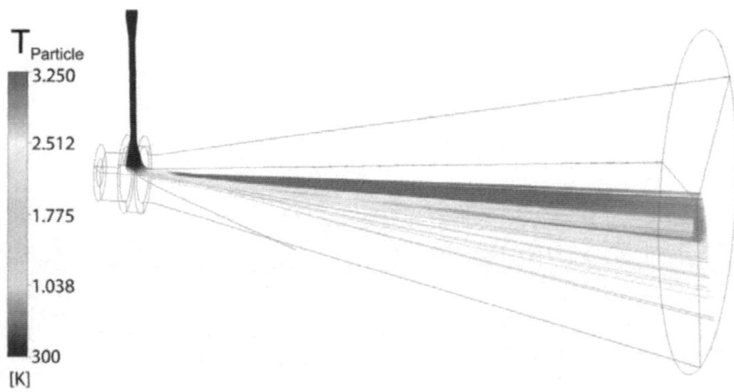

Figure 3. Temperature distribution and trajectories of the particles in the plasma and the free jet

Temperature gradients within particles were calculated exemplarily for the particles of $D = 22$ µm and $D = 60$ µm diameter and their temperature variations were compared. Gas temperatures and velocities as well as the particle velocities in the free jet at instantaneous particle coordinates were used as boundary conditions in order to calculate the temperature variation inside the particles. Temperature distributions between the particle cores and their surfaces are shown in Figure 4, where

IOP Conf. Series: Journal of Physics: Conf. Series **825** (2017) 012002 doi:10.1088/1742-6596/825/1/012002

r/R represents the particle radial coordinate normalised with respect to the initial particle radius. Since the temperature distribution of a particle in flight changes continuously, the temperature distributions are depicted at different stand-off distances y, which is the axial distance between the plasma generator nozzle and the instantaneous coordinate of the particle during its flight. For both particles, it can be seen that the particle surfaces warm up rapidly and reach high temperatures very near the plasma generator nozzle while the particle core remains cold. The temperature difference between the particle core and its surface levels out as it travels further along its flight trajectory. For the smaller particle, the particle core temperature reaches the melting temperature at $L = 17$ cm stand-off distance, whereas the core of the larger particle does not melt during the whole time of its flight.

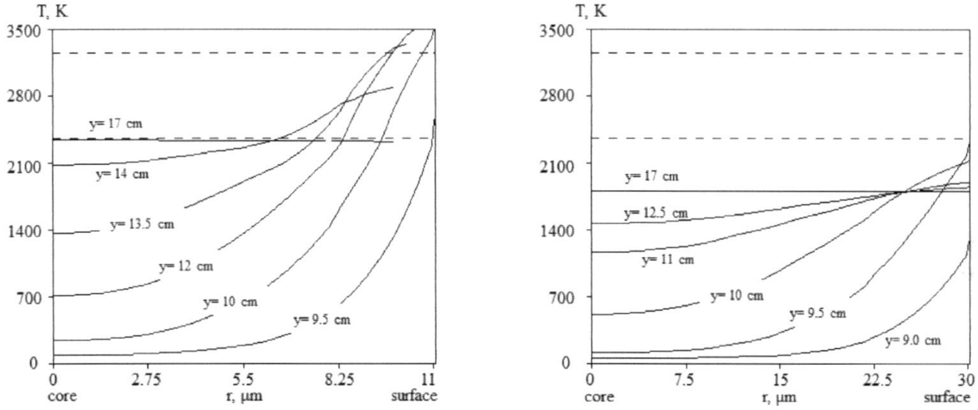

Figure 4. Variation of particle temperatures for particles of $\boldsymbol{D} = \mathbf{22}$ $\mathbf{\mu m}$ and $\boldsymbol{D} = \mathbf{60}$ $\mathbf{\mu m}$ diameter where y corresponds to the axial distance between the plasma generator nozzle and the particle in-flight position

Another way to compare small and large particles is to plot the temperature at the particle surface and its core against the distance travelled by the particle in flight. Figure 5 illustrates rapid increase of the surface temperature near the nozzle which can be explained by high plasma temperatures in the vicinity of the plasma generator nozzle. As the particles travel to the cooler regions of the free jet the surface temperatures sink and the core temperatures increase due to the thermal conductance within the particles. The temperature inside the considered particles levels out at a distance of approximately $L = 15$ cm from the nozzle, independent of the particle size. Surface and the core temperatures meet slightly above the melting point of Al_2O_3 for the smaller particle, whereas temperature is levelled out at a temperature below the melting point for the larger one. The proof of concept model developed in this work does not include the latent heat of melting. The model developed in this work has shown the significance of the heat transfer within the particle because the temperature calculated according to the lumped capacitance model holds true only starting from the $L = 15$ cm from the nozzle, which is longer than the usual spraying distance. At shorter distances the actual temperature within the particle can vary by several hundreds of degrees from the temperature determined by the lumped capacitance method.

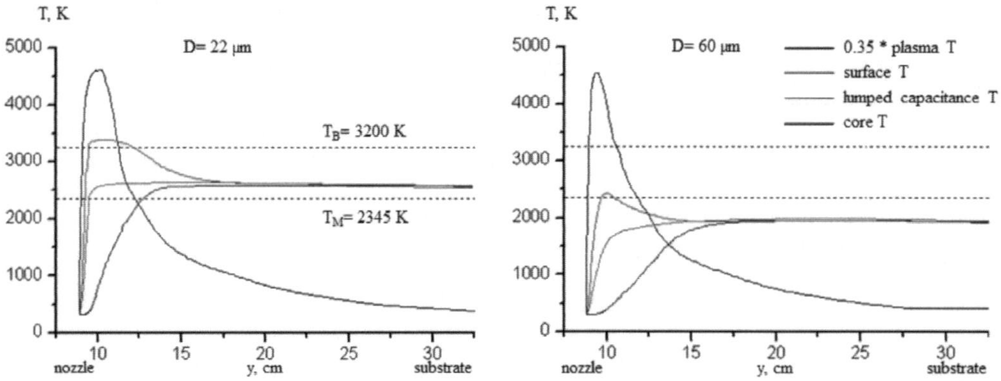

Figure 5. Temperature variation of the plasma, surface and the core of the particles as well as the particle temperature according to the lumped capacitance method during their free flight. The plasma temperature has been scaled down for better visibility

5. Summary and Outlook

In this work, the first model for the determination the temperature gradients within particles in a free jet was developed and simulated using a small and a large particle along their flight trajectory. To achieve this, three simulation models were coupled with each other. The first coupling took place between the simulation of the plasma generator and the free jet by exporting the flow profiles at the plasma generator nozzle as an inlet boundary condition for the free jet simulation. For the second coupling, gas temperatures and velocities at the immediate vicinity of the particle as well as particle velocity were used as boundary conditions for the transient heat transfer simulation within the particle along its flight trajectory. For the examined process parameters, the temperature gradients inside the small and the large particles diminish at around $\boldsymbol{L} = \mathbf{15\ cm}$ from the nozzle and the temperature of the particles become uniform. The smaller particle is completely molten at the end of the flight trajectory, whereas the core of the larger one remains solid during the whole flight. The influence of latent heat was neglected in the presented simulation. Since it is expected to be of relevance, a more sophisticated model accounting for the latent heat is in development.

6. Acknowledgements

All presented investigations were conducted in the context of the Collaborative Research Centre SFB1120 "Precision Melt Engineering" at RWTH Aachen University and funded by the German Research Foundation (DFG). For the sponsorship and the support we wish to express our sincere gratitude.

References

[1] Trelles JP, Chazelas C, Vardelle A, Heberlein JV, 2009 Arc plasma torch modeling. *Journal of Thermal Spray Technology:*728–52, 18.

[2] Fauchais P, 2004 Understanding plasma spraying. *J. Phys. D: Appl. Phys.:*R86-R108. doi:10.1088/0022-3727/37/9/R02, 37.

[3] Baudry C, Vardelle A, Mariaux G, Delalondre C, Meillot E, Three-dimensional and time-dependent model of the dynamic behavior of the arc in a plasma spray torch. In: ITSC 2004: International Thermal Spray Conference 2004: Advances in Technology and Application.

[4] Chazelas C, Moreau E, Mariaux G, Vardelle A, 2006 Numerical modeling of arc behavior in a DC plasma torch. *High Temperature Material Processes: An International Quarterly of High-Technology Plasma Processes*, 10.

[5] Bobzin K, Öte M, 2015, A Numerical Investigation: Air Plasma Spraying by means of a Three-Cathode Spraying Torch. In: International Thermal Spray Conference and Exposition (ITSC).

[6] Bobzin K, Öte M, 2016 Modeling Multi-Arc Spraying Systems. *J Therm Spray Tech:*920–32. doi:10.1007/s11666-016-0407-7, 25.

[7] Bobzin K, Kopp N, Warda T, Schäfer M, Öte M A Numerical Investigation: Influence of the Operating Gas On the Flow Characteristics of a Three-Cathode Air Plasma Spraying System. In: International Thermal Spray Conference and Exposition (ITSC).

[8] Bobzin K, Öte M, Schein J, Zimmermann S, Möhwald K, Lummer C, 2016 Modelling the Plasma Jet in Multi-Arc Plasma Spraying. *Journal of Thermal Spray Technology:*1111–26, 25.

[9] Vardelle M, Vardelle A, Leger A. C, Fauchais P, Gobin D, 1995 Influence of particle parameters at impact on splat formation and solidification in plasma spraying processes, *Journal of Thermal Spray Technology:* 50-8, 4. doi:10.1007/BF02648528

[10] Pasandideh-Fard M, Pershin V, Chandra S, Mostaghimi J, 2002 Splat shapes in a thermal spray coating process: Simulations and experiments, *Journal of Thermal Spray Technology:* 206-17, 11. doi:10.1007/BF02648528

[11] Khelfi D, El-Hadj AA, Aït-Messaoudène N, 2008 Modeling of a 3D plasma thermal spraying and the effect of the particle injection angle, *Revue des Energies Renouvelables CISM'08:* 205-16.

[12] Bobzin K, 2013 Oberflächentechnik für den Maschienenbau. Germany: Weinheim Wiley-VCH Verlag.

[13] Ahmed I, Bergman T. L, 2000 Three-dimensional simulation of thermal plasma spraying of partially molten ceramic agglomerates, 215-24, 9. doi:10.1361/105996300770349953

[14] McKelliget JW, Trapaga G, Gutierrez-Miravete E, Cybulski M, 1998 An integrated mathematical model of the plasma spraying process, *Thermal Spray: Meeting the Challenges of the 21st Century:* 335-40, 1.

[15] Chen X, Chyou YP, Lee YC, Pfender E, 1985 Heat Transfer to a Particle under Plasma Conditions with Vapor Contamination from the Particle, *Plasma Chemistry and Plasma Processing*: 119-41, 5.

[16] Wan YP, Prasad V, Wang G-X, Sampath S, Fincke JR, 1999 Model and Powder Particle Heating, Melting, Resolidification, and Evaporation in Plasma Spraying Processes, *Journal of Heat Transfer*: 691-9, 121.

[17] Li M, Shi D, Christofides PD, 2005 Modeling and control of HVOF thermal spray processing of WC–Co coatings, *Powder Technology:* 177-94, 156.

[18] Saha K, Chaudhuri S, Cetegen BM, 2011 Modeling of Ceramic Particle Heating and Melting in a Microwave Plasma, *Journal of Heat Transfer:* 133. doi: 10.1115/1.4002448

[19] Dyshlovenko S, Pawlowski L, Pateyron B, Smurov I, Harding JH, 2005 Modelling of plasma particle interactions and coating growth for plasma spraying of hydroxyapatite, *Surface and Coatings Technology:* 3757-69, 200.

[20] Zhang W, 2008 Integration of Process Diagnostics and Three Dimensional Simulations in Thermal Spraying; Dissertation.

[21] Fauchais P, Vardelle M, 2010 Sensors in Spray Processes. *Journal of Thermal Spray Technology:*668–94.

[22] Bauder U, 1968 Radiation from high - pressure plasmas. *Journal of Applied Physics:*148–52, 39.

[23] Menart J, Malik S 2002 Net emission coefficients for argon-iron thermal plasmas. *J. Phys. D: Appl. Phys.:*867–74, 35. doi:10.1088/0022-3727/35/9/306.

[24] Murphy AB, Arundelli CJ, 1994 Transport coefficients of argon, nitrogen, oxygen, argon-nitrogen, and argon-oxygen plasmas. *Plasma Chemistry and Plasma Processing:* 451–90, 14.

[25] Murphy AB, Tam E, 2014 Thermodynamic properties and transport coefficients of arc lamp plasmas: argon, krypton and xenon. *Journal of Physics D: Applied Physics*: 47.29

[26] Borisov Y, Bushma A, Krivtsun I, 2006 Modeling of motion and heating of powder particles in laser, plasma, and hybrid spraying. *Journal of Thermal Spray Technology:*553–8, 15.

[27] Öte M, 2016 Understanding Multi-Arc Plasma Spraying. 1. Auflage. Herzogenrath: Shaker, Dissertation ISBN 978-3-8440-4598-7

[28] Lamoreaux RH, Hildenbrand DL, Brewer L, 1987 High - Temperature Vaporization Behavior of Oxides II. Oxides of Be, Mg, Ca, Sr, Ba, B, Al, Ga, In, Tl, Si, Ge, Sn, Pb, Zn, Cd, and Hg. *Journal of physical and chemical reference data:*419–43, 16.

Numerical Coupling of the Particulate Phase to the Plasma Phase in Modeling of Multi-Arc Plasma Spraying

K. Bobzin, M. Öte*
RWTH Aachen University, IOT - Surface Engineering Institute, Aachen, Germany

E-Mail: oete@iot.rwth-aachen.de, *corresponding Author

Abstract. Inherent to Euler-Lagrange formulation, which can be used in order to describe the particle behavior in plasma spraying, particle in-flight characteristics are determined by calculating the impulse, heat and mass transfer between the plasma jet and individual powder particles. Based on the assumption that the influence of the particulate phase on the fluid phase is insignificant, impulse, heat and mass transfer from particles to the plasma jet can be neglected using the so-called numerical approach of "one-way coupling". On the other hand, so-called "two-way coupling" considers the two-sided transfer between both phases. The former is a common simplification used in the literature to describe the plasma-particle interaction in thermal spraying. This study focuses on the significance of this simplification on the calculated results and shows that the use of this simplification leads to significant errors in calculated plasma and particle in-flight characteristics in three-cathode plasma spraying process.

1. Introduction

Coating formation in plasma spraying takes place as a result of continuous deposition of the molten and semi-molten particles on to the workpiece surface. Particle in-flight characteristics, i.e. their velocities, temperatures, diameters and melting grades, determine the coating quality and deposition efficiency. Particle in-flight characteristics themselves are determined by the impulse, heat and mass transfer between the plasma jet and the powder particles.

Several numerical models have been developed in the last decades, focusing on the plasma jet-particle interaction in the single-arc plasma spraying process. These have all contributed to an in-depth understanding of the impulse and heat transfer onto the particles during their residence in the plasma jet. Several models with different abstraction levels have been developed and great progress has been achieved in this field [1-12].

All previous works which have dealt with particle-plasma interaction in plasma spraying have in common that the authors have employed different model simplifications to explain certain aspects, mostly dealing with so-called non-transferred conventional single-arc plasma generators. A comprehensive numerical research focusing on plasma-particle interaction in the case of new generation multi-arc systems has not been conducted yet. One of the major assumptions employed in numerical works done so far is that the influence of particles on plasma jet characteristics is negligible in plasma spraying. The aim of this study is therefore to investigate this effect and identify the validity of the above-mentioned assumption. Meanwhile, this study focusses on multi-arc plasma spraying of ceramic feedstock materials.

2. State of Art

Heat and mass transfer between the plasma jet and the injected powder particles can principally be calculated using the Euler-Euler approach or the Euler-Lagrange approach. In both cases, a distinction is made between the continuous (plasma gas) and the dispersed phase (particles). The Euler-Euler approach treats both phases as a continuum in which they can penetrate and influence each other. The Euler-Lagrange approach on the other hand describes only the plasma gas as a continuum, whereas the particles are described as a discrete phase. By calculating the momentum transfer acting on the particles, the trajectories can be predicted and calculated. Similarly, by calculating the heat transfer, the temperatures of the particles can be determined. Due to its simplicity and robustness, the Euler-Lagrange approach dominates the literature dealing with the modelling of plasma-particle interaction in thermal spraying.

The momentum transfer in a plasma jet is influenced by several forces including virtual drag force, virtual mass force, pressure gradient force, basset force, turbulent dispersion force and external forces such as electric, magnetic and gravitational forces [13]. Although the influences of different types of forces might vary in significance depending on the powder and feed stock material characteristics as well as on injection and process conditions, all forces except the viscous drag play a secondary role in the momentum transfer between plasma jet and powder particles. This might be attributed to the fact that the densities of the feedstock materials in plasma spraying are significantly higher than that of the plasma gas. Please refer to [13] for further discussion regarding the significance of different force terms on the momentum transfer.

In the broadest sense, heat transfer between plasma jet and particles takes place as a result of convection, radiation and mass transfer. Convective heat transfer is generally considered to be more significant compared to other heat transfer mechanisms. Investigations show that although the heat transfer due to radiation gains importance with increasing spraying distance, the overall heat transfer due to convection is usually two orders of magnitude higher in comparison to radiation [14].

Euler-Lagrange approach requires the determination of the drag and heat transfer coefficients between the species of the flow for the calculation of momentum transfer due to drag forces and for the calculation of heat transfer due to convection. The equations used to determine these coefficients in case of ordinary flow conditions need to be further modified in order to make them suitable for the flow conditions in thermal spraying. In the literature, different correction factors considering the particularities of thermal plasma jets have been proposed for this purpose. For example, steep temperature gradients across the boundary layers surrounding the single particles cause the temperature dependent transport and thermodynamic properties of the plasma gas to change drastically across this layer. Determining the drag and heat transfer coefficients based on the properties of the plasma gas outside this layer would lead to erroneous results. Therefore, adjustments by means of a correction factors have been employed in the literature in order to take this effect into consideration [15-20]. Similarly, the Knudsen effect and basset history forces have been treated using correction factors [13, 17, 20-23], when required.

Mass transfer between particles and the plasma jet can occur as a result of evaporation of molten droplets or due to the chemical reaction between the particle phase and plasma gas constituents, i.e. due to oxidation. The physical evaporation of the feedstock material can be modeled using simple heat transfer models by controlling the vaporization rate with heat flux divided by latent heat of vaporization [24, 25]. On the other hand, vaporization due to volatile species requires more complicated models. Only in a few works, the numerical modeling of this phenomenon in thermal spraying is reported [12, 15, 26]. In their respective work, Wan et al. give a good overview of these activities [27].

Previous works which have dealt with the particle-plasma interaction in thermal spraying [1-12] have one thing in common: the authors have employed different model simplifications to explain certain aspects. A discussion about these simplifications, their underlying assumptions or their significance goes beyond the scope of this paper. This paper focusses on the common assumption that the influence of particles on plasma jet characteristics is negligible in plasma spraying and deals with the influence of corresponding model simplification to the results. None of the above-mentioned works has dealt with the significance of numerical back-coupling of powder particles, which allows the particles to influence the fluid phase via source terms of heat, momentum and mass on the results.

3. Mathematical Model and Boundary Conditions

Plasma exits the plasma generator at the nozzle outlet at high temperature and velocity. The set of equations which are used for modeling the plasma jet outside the plasma torch corresponds to the set of equations used for plasma generator simulations with the exception of the equations describing electromagnetic phenomena. For an overview, please refer to [28, 29]. The Euler-Lagrange approach is used in this work in order to study the particle behavior in plasma jet. The calculation domain involves the region downstream of the torch outlet. The boundary conditions and the calculation domain used in this study are presented in Figure 1.

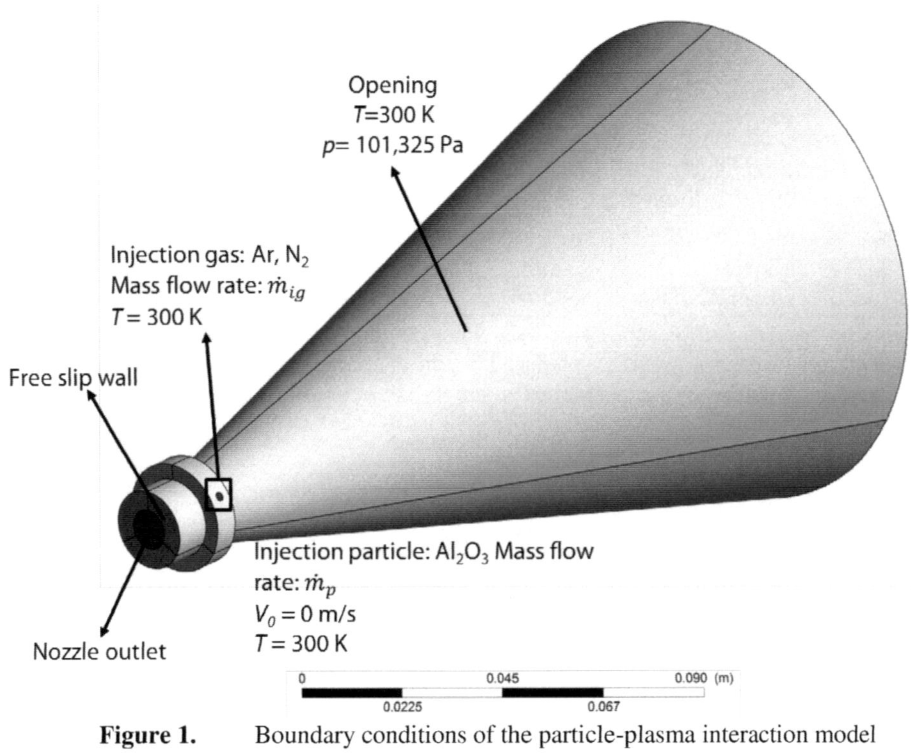

Figure 1. Boundary conditions of the particle-plasma interaction model

The boundary conditions at the nozzle outlet are imported from a-priori conducted plasma generator simulations. Details of the model which is used to describe the physical processes in a three-cathode plasma generator have already been published in [30]. The developed model dealing with the flow characteristics of the plasma jet is introduced in [31]. The opening boundary condition at the outer surface of the calculation domain represents the flow behavior in the infinity of the air atmosphere. Moreover, an inlet boundary condition has been defined, over which powder particles are injected in the calculation domain with defined mass flow rates, velocities and size distributions. The material data used in the simulations are listed in Table 1. The specific heat capacity of the feedstock material is defined as temperature and physical state-dependent using NASA polynomials. Values of latent heat of fusion and boiling are calculated using the enthalpies of the corresponding states at melting and boiling temperatures. The enthalpy values are calculated once again using NASA polynomials [32]. Although Al_2O_3 exhibits higher thermal conductivities of about κ = 37 W/mK at room temperature, this value decreases rapidly down to κ = 5-6 W/mK at temperatures higher than T = 1,200 K [33, 34]. In this work, for the sake of simplicity, a constant value of thermal conductivity, which is corresponding to the value of it at higher temperatures, is used.

Table 1: Material data used in the plasma-particle interaction simulations

Al_2O_3	Solid state	Liquid phase	Gas phase
Density [g/cm^3]	3.95	2.7	0.001[a]
Thermal conductivity [W/mK]	6	6	-
Specific heat capacity [J/kgK]	~600-900[b]	~1950[b]	~1200[b]
Latent heat of fusion (kJ/kg)		1060	
Latent heat of boiling (kJ/kg)		8885	
Melting point [K]		2327	
Boiling point [K]		3250	

[a] assumed value
[b] defined temperature dependent in the models

The particle motion in plasma spraying as a result of the drag force (F_D) is described according to equation (1). The value of the drag coefficient (C_D) in this equation depends on the particle's Reynolds number (Re_p) and forms the basis of the momentum transfer calculations. The correlation between the drag coefficient and the particle's Reynolds number employed in this study corresponds to the correlation proposed by Vardelle et al. in [20], as depicted in equation (2). Re_p itself is a function of density (ρ_g) and dynamic viscosity (μ_g) of plasma gas, particle diameter (d_p) and slip velocity between particle and plasma gas (s) as expressed in equation (3). In this study, the varying properties of the plasma gas across the boundary layer of the particle are accounted for using a simple averaging conducted over five temperature points lying between plasma gas (T_g) and particle temperature (T_p)

$$F_D = m_p \frac{dv_p}{dt} = \frac{1}{2} C_D A_p \rho_g |v_p - v_g| (v_g - v_p) \tag{1}$$

$$C_D = \begin{cases} \dfrac{24}{Re_p}, & Re_p \leq 0.2 \\ \dfrac{24}{Re_p}(1 + 0.1 Re_p^{0.99}), & 0.2 < Re_p < 2 \\ \dfrac{24}{Re_p}(1 + 0.11 Re_p^{0.81}), & 2 < Re_p < 21 \\ \dfrac{24}{Re_p}(1 + 0.189 Re_p^{0.62}), & 21 < Re_p \leq 500 \\ 0.44, & 500 < Re_p \end{cases} \tag{2}$$

IOP Conf. Series: Journal of Physics: Conf. Series **825** (2017) 012003 doi:10.1088/1742-6596/825/1/012003

$$Re_p = \frac{\rho_g s\, d_p}{\mu_g} \quad (3)$$

The convective heat transfer (Q_c) between plasma jet and particles is described according to equation (4). In this equation, h stands for the heat transfer coefficient. It is often convenient to express the heat transfer coefficient in terms of a dimensionless Nusselt number (Nu_p) as it is shown in equation (5). For a particle moving in plasma jet, Nu_p is a function of the particle's Reynolds number (Re_p) and gas Prandtl number (Pr_g). The correlation between Nu_p, Re_p and Pr_g employed in this study corresponds to the correlation proposed by Hughmark in [35], as depicted in equation (6). Pr_g is a function of specific heat capacity (Cp_g), dynamic viscosity (μ_g) and thermal conductivity (κ_g) of plasma gas according to the relationship expressed in equation (7). Analog to the situation for the drag coefficient, the varying plasma gas properties across the boundary layer of the particle are accounted for using a simple averaging.

$$Q_c = hA_i(T_g - T_p) \quad (4)$$

$$h = \frac{\kappa_g Nu_p}{d_p} \quad (5)$$

$$Nu_p = \begin{cases} 2 + 0.60 Re_p^{0.50} Pr_g^{0.33}, & 0 \le Re_p < 776.06 \text{ and } 0 \le Pr_g < 250 \\ 2 + 0.27 Re_p^{0.62} Pr_g^{0.33}, & 776.06 \le Re_p \text{ and } 0 \le Pr_g < 250 \end{cases} \quad (6)$$

$$Pr_g = \frac{Cp_g \mu_g}{\kappa_g} \quad (7)$$

In order to calculate the phase change without resolving the temperature distribution within the particle, a simple mathematical formulation is proposed and used. The melting rate ($\dot{m}_{s\to l}$) in kg/s is correlated with the heat transfer rate between plasma gas and particle (1st term) as well as with the averaged conductive heat transfer rate within the particle (2nd term), according to the equation (8):

$$\dot{m}_{s\to l} = \begin{cases} \dfrac{\pi \kappa_g Nu_p d_p (T_g - T_p)}{L_f} + \dfrac{\pi \kappa_p d_p (T_p - T_{m,p})}{L_f}, & T_{m,p} < T_p \\ 0, & T_{m,p} > T_p \end{cases} \quad (8)$$

Equation (8) describes mathematically that the rate of the phase change is determined by the heat transfer rate divided by the latent heat of fusion (L_f). Phase change does not take place at particle temperatures below the melting point of the feedstock material ($T_{m,p}$). By coupling the energy sink term ($\dot{m}_{s\to l} \cdot L_f$) into the energy equations of the gas flow, the implementation of the mathematical formulation into the model is closed. Please note that, inherent to the characteristics of the Lagrangian formulation, the particle temperatures (T_p) in equation (8) correspond to "lumped" particle temperatures. This means that no distinction between the surface temperatures and temperatures within the particle is made. Moreover, the developed model accounts for the re-solidification of the particles, too. A similar approach underlying equation (8) has been employed for this purpose. The energy released due to re-solidification has in this case been coupled into the energy equations of plasma gas flow as an energy source term.

Mass transfer between particles and plasma jet due to physical evaporation is described according to equation (9). In this equation, $\dot{m}_{l\rightarrow g}$ stands for the rate of evaporation. $T_{b,p}$ and L_b are boiling point and latent heat of boiling for the feedstock material. The mathematical formulation is coupled with the energy equations through the energy sink term ($\dot{m}_{l\rightarrow g} \cdot L_b$). The resulting particle vapor is transmitted to the Euler phase via the mass source term ($\dot{m}_{l\rightarrow g}$). Other mass transfer mechanisms, for example due to emergence of some persistent or volatile reaction products, are not taken into consideration.

$$\dot{m}_{l\rightarrow g} = \begin{cases} \dfrac{\pi \kappa_g N u_p d_p (T_g - T_p)}{L_b}, & T_{b,p} < T_p \\ 0, & T_{b,p} > T_p \end{cases} \tag{9}$$

4. Results and Discussion

In case of the so-called numerical approach of "one-way coupling", only the influence of the fluid phase on the particulate phase is considered and the influence of the particulate phase on the fluid phase is neglected. On the other hand, "two-way coupling" allows the particles to influence the fluid phase via source terms of heat, momentum and mass. The former is a common simplification used in literature and is justified with the rate of particle injection being too low in plasma spraying to influence the plasma jet [12].

The model introduced in chapter 3 describes the behavior of ceramic powder particles in a plasma jet generated by means of a three-cathode plasma generator. Accounting the typical size distribution and material properties of the feedstock material Al_2O_3, some of the momentum and heat transfer mechanisms have been neglected. The model developed in this study takes almost all relevant effects into account, including melting, re-solidification and physical evaporation of the particles. Merely evaporation as a result of decomposition of the ceramic feedstock materials below the boiling point is not taken into consideration, the relevance of which is not investigated and clarified completely yet. Therefore, the model offers a good basis in order to analyze the differences resulting from "one-way" and "two-way coupling" approaches.

In Figure 2, the results of the simulations conducted with "one-way" and "two-way coupling" are illustrated. The comparison is conducted for a particle mass flow rate of $\dot{m}_p$ = 24 g/min per injector. This is a typical rate for ceramic feedstock materials employed in thermal spraying. Powder which has an assumed normal particle size distribution with a mean value of μ = 30 µm and a standard deviation of σ = 10 µm is injected from the same position in both cases. All the other parameters are kept constant.

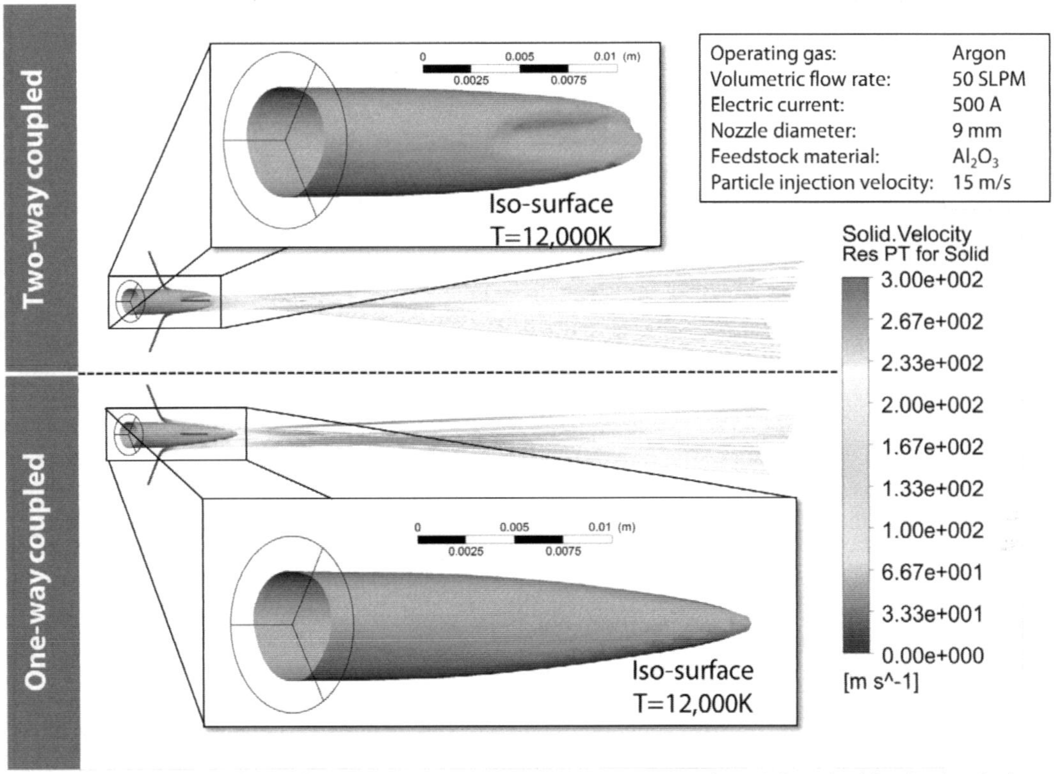

Figure 2. Comparison of calculated plasma gas temperatures and particle velocities in simulations which employ one-way and two-way coupling [36].

The results show that the particle injection clearly reduces the plasma temperatures leading to a slightly shorter plasma jet length. Moreover, it is apparent that in the injection region of the powder, the iso-surface representing the plasma temperatures at $T = 12{,}000$ K exhibits a distorted profile. This implies that the injected powder particles cool down the plasma jet in this region significantly. Neglecting the reduction of the plasma jet temperatures due to this effect causes in turn that the calculations of the particle in-flight characteristics show significant errors. On the one hand, the calculated heat and mass transfer between plasma gas and particles as well as melting degrees of particles are directly affected from over-estimated plasma gas temperatures as depicted in equations (4), (8) and (9). On the other hand, the changing thermodynamic and transport properties of the plasma gas with over-estimated plasma gas temperatures indirectly influence the calculated impulse, heat and mass transfer between plasma gas and particles. As a result, particle velocities, temperatures and melting degrees calculated with the approaches "one-way" and "two-way coupling" exhibit significant differences. Figure 2 shows exemplarily the comparison of particle velocities and trajectories which are calculated with these approaches. As it is evident that the plasma jet and particle in-flight characteristics are significantly influenced by the injected particles, the one-way coupling simplification shall be avoided for an accurate description of the plasma-particle interaction in plasma spraying.

5. Conclusion

A comprehensive numerical model focusing on plasma-particle interaction in case of new generation multi-arc plasma spraying systems has not been developed in the literature yet. In this study, this has been realized on the example of plasma jet generated by a three-cathode plasma spraying system. The developed model takes almost all relevant effects into account, including melting, re-solidification and physical evaporation of the particles. On the basis of this model, the significance of the often used assumption that the particle phase does not have a significant influence on the plasma jet characteristics is critically evaluated. It has been shown that this assumption leads to significant errors in calculated plasma and particle in-flight characteristics. Therefore, the so-called "one-way coupling" approach, which is based on the above-mentioned assumption, shall be avoided for an accurate numerical description of the plasma-particle interaction in plasma spraying.

Acknowledgements

All presented investigations were conducted in the context of the Collaborative Research Centre SFB1120 "Precision Melt Engineering" at RWTH Aachen University and funded by the German Research Foundation (DFG). For the sponsorship and the support, we wish to express our sincere gratitude. This paper is partially based on chapter 4 in M. Öte, ''Understanding Multi-Arc Plasma Spraying,'' Shaker-Verlag, RWTH Aachen, Dissertation, ISBN: 978-3-8440-4598-7.

References

[1] Bobzin K, Kopp N, Warda T, Petkovic I, Zimmermann S, Hartz-Behrend K, Landes K, Forster G, Kirner S, Marqués J-L, Schein J, Prehm J, Möhwald K and Bach F W 2013 Improvement of Coating Properties in Three-Cathode Atmospheric Plasma Spraying *Journal of Thermal Spray Technology* **22**(4) pp 502-8

[2] Kang C W, Ng H W, Yu S C M 2007 Plasma Spray Deposition on Inclined Substrates: Simulations and Experiments *Journal of Thermal Spray Technology* **16**(2) pp 261-74

[3] Li H-P, Pfender E 2007 Three Dimensional Modeling of the Plasma Spray Process *Journal of Thermal Spray Technology* **16**(2) pp 245-60

[4] Meillot E, Balmigere G 2008 Plasma spraying modeling: particle injection in a time-fluctuating plasma jet *Surface and Coatings Technology* **202**(18) pp 4465-69

[5] Petkovic I 2014 Modellierung und Simulation als Werkzeug zur Verbesserung des APS-Prozesses mittels Mehrlichtbogen-Plasmagenerator *Shaker-Verlag* (RWTH Aachen Dissertation) pp 1-192

[6] Planche M P, Bolot R, Coddet C 2003 In-flight characteristics of plasma sprayed alumina particles: measurements, modeling, and comparison *Journal of Thermal Spray Technology* **12**(1) pp 101-11

[7] Thiyagarajan T K, Sreekumar K P, Selvan V, Ramachandran K, Ananthapadmanabhan P V 2010 Simulation studies to optimize the process of plasma spray deposition of yttrium oxide *Journal of Physics Conference Series* **208**(1)

[8] Xu D-Y, W X-C, Chen X 2003 Motion and heating of non-spherical particles in a plasma jet *Surface and Coatings Technology* **171**(1–3) pp 149-56

[9] Vardelle M, Vardelle A, Fauchais P, Li K-I, Dussoubs B, Themelis N J 2001 Controlling particle injection in plasma spraying *Journal of Thermal Spray Technology* **10**(2) pp 267-84

[10] Zhang T, Gawne D T, Liu B 2000 Computer modelling of the influence of process parameters on the heating and acceleration of particles during plasma spraying *Surface and Coatings Technology* **132**(2–3) pp 233–43

[11] Zhang W, Zheng L L, Zhang H, Sampath S 2007 Study of Injection Angle and Carrier Gas Flow Rate Effects on Particles In-Flight Characteristics in Plasma Spray Process: Modeling and Experiments *Plasma Chemistry and Plasma Processing* **6** pp 701-16

[12] Westhoff R, Trapaga G, Szekely J 1992 Plasma-particle interactions in plasma spraying systems *Metallurgical and Materials Transactions B* **23** pp 683-93

[13] Pfender E, Lee Y C 1985 Particle Dynamics and Particle Heat and Mass Transfer in Thermal Plasmas, Part I: the Motion of a Single Particle without Thermal Effects *Plasma Chemistr and Plasma Processing* **5**(3) pp 211-37

[14] Wan Y P, Prasad V, Wang G-X, Sampath S, Fincke J R 1999 Model and Powder Particle Heating, Melting, Resolidification, and Evoporation in Plasma Spraying Processes *Journal of Heat Transfer* **121** pp 691-99

[15] Fiszdon J K 1979 Melting of powder grains in a plasma flame *International Journal of Heat Mass Transfer* **22**(5) pp 749-61

[16] Lee Y C, Chyou Y P, Pfender E 1987 Particle Dynamics and Particle Heat and Mass Transfer in Thermal Plasmas, Part II: Particle Heat and Mass Transfer in Thermal Plasmas *Plasma Chemistry and Plasma Processing* **5**(4) pp 391-414

[17] Lewis J A, Gauvin W H 1973 Motion of particles entrained in a plasma jet *AIChE Journal* **19**(5) pp 982-90

[18] Sayegh N, Gauvin W 1979 Numerical analysis of variable property heat transfer to a single sphere in high temperature surroundings *AIChE Journal* **25**(3) pp 523–34

[19] Young R M, Pfender E 1987 Nusselt number correlations for heat transfer to small spheres in thermal plasma flows *Plasma Chemistry and Plasma Processing* **7**(2) pp 211-29

[20] Vardelle M, Vardelle A, Fauchais P, Boulos M I 1983 Plasma-particle momentum and heat transfer: Modelling and measurements *AIChE Journal* **9**(2) pp 236–43

[21] Bhattacharya D, Gauvin W H 1975 Modeling of heterogeneous systems in a plasma jet reactor *AIChE Journal* (Hoboken, NJ) **21**, pp 879-85

[22] Chen X, Pfender E 1983 Effect of the Knudsen Number on Heat Transfer to a Particle Immersed into a Thermal Plasma *Plasma Chemistry and Plasma Processing* **3**(1) p 97

[23] Fauchais P 2004 Topical Review: Unterstanding plasma spraying *Journal of Physics D. Applied Physics* **37**(9) pp 86-108

[24] Ptoulx P, Mostaghimi J, Boulos M I 1985 Plasma-particle Interaction Effects in Induction Plasma Modeling under Dense Loading Conditions *International Journal of Heat and Mass Transfer* **28**(7) pp 1327-36

[25] Das D K, Sivakumar R 1990 Modeling of the Temperature and the Velocity of Ceramic Powder Particles in A Plasma Flame-I. Alumina *Acta Metallurgica et Materialia* **38**(11) pp 2187-92

[26] Vardelle A, Themelis N J, Dussoubs B, Vardelle M, Fauchais P 1997 Transport and chemical rate phenomena in plasma sprays *High Temperature Material Processes* **1** pp 295-314

[27] Wan Y P, Fincke J R, Sampath S, Prasad V, Herman H 2002 Modeling and experimental observation of evaporation from oxidizing molybdenum particles entrained in a thermal plasma jet *International Journal of Heat and Mass Transfer* **45**(5), pp 1007–15

[28] Bobzin K, Bagcivan N, Zhao L, Petkovic I, Schein J, Hartz-Behrend K, Kirner S, Marqués J-L, Forter G 2011 Modelling and diagnostics of multiple cathodes plasma torch system for plasma spraying *Frontiers of Mechanical Engineering* **6** p 3

[29] Bobzin K, Bagcivan N, Petkovic I 2011 Numerical and experimental determination of plasma temperature during air plasma spraying with a multiple cathodes torch *Journal of Materials Processing Technology* **211**(10) pp 1620-28

[30] Bobzin K, Öte M 2016 Modelling Multi-Arc Spraying Systems *Journal of Thermal Spray Technology* **25**(5) pp 920-32

[31] Bobzin K, Öte M, Schein J, Zimmermann S, Möhwald K, Lummer C 2016 Modelling the Plasma Jet in Multi-Arc Plasma Spraying *Journal of Thermal Spray Technology* **25**(6) pp 1111-26

[32] Burcat A, Ruscic B 2005 Third Millenium Ideal Gas and Condensed Phase Thermochemical Database for Combustion with Updates from Active Thermochemical Tables *Argonne National Laboratory* pp 1-417

[33] Abdou M A, Tillack M S, Raffray A R, Hadid A H, Bartlit J R, Bell C E C, Gierszewski P J, Gordon J D, Iizuka T, Kim C N 1988 Modeling, analysis and experiments for fusion nuclear technology *FNT Progress Report: Modeling and FINESSE* (L.A.: University of California in L.A., UCLA-ENG-86-44)

[34] Touloukian Y S 1966 Thermophysical Properties of High Temperature Solid Materials. Volume 4. Oxides and Their Solutions and Mixtures. Part 1. Simple Oxygen compounds and Their Mixtures *Thermophysical and Electronic Properties Information Analysis Center*

[35] Hughmark G A 1967 Mass and heat transfer from rigid spheres *AIChE Journal* **13**(6) pp 1219-21

[36] Bobzin K, Öte M 2016 Influence of Powder Particles on the Plasma Characteristics in Multi-arc Plasma Spraying *Proceedings of 14th High-Tech Plasma Processes Conference HTPP 14* (Münich) p 89

DBD tranformerless power supplies: impact of the parasitic capacitances on the power transfer.

M A Diop[1], A Belinger[1], H Piquet[1]

[1]LAPLACE, Université de Toulouse, CNRS, INPT, UPS, France

belinger@laplace.univ-tlse.fr

Abstract. A new transformerless power supply for DBD application is presented here. The power supply is built with *10kV* SiC MOSFET. This high voltage switches allow holding the high voltage required by the DBD. An analytical study of the converter's operation is presented to deduce the power transmitted to the DBD. A comparison between the experimental and theoretical electrical waveforms is shown. The experimental waveforms are particularly affected by all the parasitic capacitances. When all the switches are in OFF state, oscillations cause over-voltages across the switches. An analysis of the effect of each capacitance is presented and demonstrates that the parasitic capacitances of the switches and of the inductance play a key role in the actual power transfer.

1. Introduction

Dielectric Barrier Discharges (DBD) is a device where it is easy to establish cold plasma at atmospheric pressure [1]. For this reason, DBDs have a lot of applications: ozone generator [2], excimer lamp [3], surface treatment [4], flow control actuators [5]... In Dielectric Barrier Discharge (DBD) setups, at least one dielectric is placed between the two metal electrodes that supply the gas. The dielectric enables to limit the local current of the discharge, even despite a large *pd* product (*p* is the pressure and *d* the gas gap). From an electrical point of view, DBD are capacitive loads. So they must be supplied by an alternative current source. Furthermore, to ignite the discharge at atmospheric pressure, the applied voltage usually reaches several kilovolts. This high alternating voltage is traditionally obtained via amplification, using a step-up transformer. A DBD is a capacitive charge: for a sinusoidal source the current and the voltage are delayed by *90* degree; this implies a high consumption of reactive power and an over-sizing of the transformer in order to sustain the maximum voltage and rms current required by the DBD. Therefore the transformer is often very heavy and cumbersome. For example for a *100W* DBD, the volume of the transformer is often up to 1 liter. In classical electronic application this volume is used for a *1kW* system. Moreover, the parasitic elements of the transformer (leakage inductance, inter-turn and inter-winding parasitic capacitance), specific to the physical structure of this device, limit the power transfer. For example, parasitic capacitances of the transformer have the same order of magnitude as those of the DBD. They are parallel to the DBD, so they divert a significant part of the current [6]. This also slows down the voltage rise and delays the ignition of the discharge.

Improving these issues and the over-sizing incited us to investigate power supplies for DBD without transformer. A high-voltage inverter is directly connected to the DBD. It involves the use of high-

IOP Conf. Series: Journal of Physics: Conf. Series **825** (2017) 012004 doi:10.1088/1742-6596/825/1/012004

voltage switches. Because high-voltage switches are available only for a few years, only theoretical developments have been proposed to date [7]. Using *10kV* SiC DMOSFET, an innovative transformerless pulsed-current source topology is proposed here and is detailed. The theoretical calculation of the transferred power is presented in a first part. Afterwards we show that the parasitic capacitances affect the operating conditions. We analyze the impact of each parasitic capacitance on the quality of the power transfer. Each capacitance is studied separately in order to understand its impact.

2. Experimental setup and power supplies operating

In a power-supply design approach, the modelling of the DBD is developed below [8]. The DBD used for the experiments is an excimer UV lamp with a cylindrical configuration, with two dielectric barriers and two coaxial tubes of quartz containing a volume of a Xe-Cl gas mixture at *120mBar* (Figure 1.a). Metal electrodes are placed on their outer walls. Concerning the geometry of the lamp, the length of the lamp is 115mm and the gap is 20mm.

As long as the discharge is in *OFF* state, the DBD is modelled by two capacitors in series: C_{diel} and C_{gas}. The latters correspond respectively to the dielectrics and to the gas capacitance. The geometrical characteristics allow to estimate these capacitances: $C_{gas}=28pF$ and $C_{diel}=85pF$. When the discharge is in *ON* state, we consider that the gas voltage remains constant, $\pm V_{th}$ (the sign depends on the direction of the current). This assumption is accurate when the discharge is a Townsend or glow mode; it does not take local phenomena into account, but concerns a macroscopic point of view. For a power-supply design approach, it is sufficient to take into account the non-linearity of the plasma [9]. As shown in Figure 1.b the discharge is electrically modelled by a (V_{gas}, I_{gas}) dipole taking into consideration the non-linearity explained above; this dipole is parallel to the gas capacitance C_{gas}. The breakdown voltage was measured at $V_{th}= 1310kV$. In the whole paper, we develop the analysis with this lamp; however the general conclusions are valid for any DBD device.

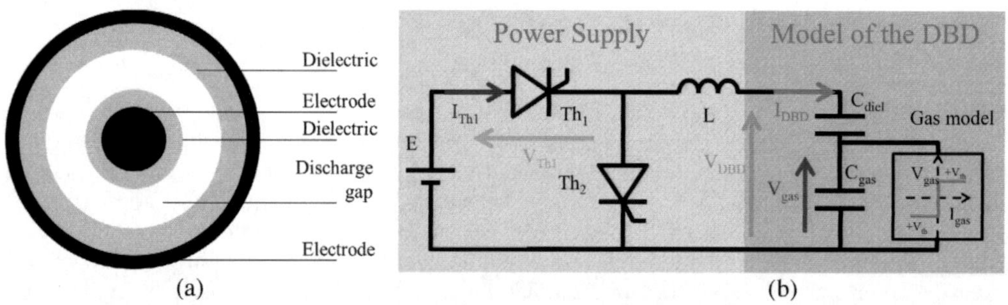

Figure 1: (a) Diagram of the DBD lamp. (b) Diagram of the power supply and electrical model of the DBD

Due to the capacitive nature of the DBD and the electrical characteristic of the discharge, the DBD must be supplied by a current source generator [10]. In steady state, the DBD current must have a null average value. Therefore, in the topology presented in Figure 1.b, an inductor is placed in series with the DBD; a constant source voltage E supplies this circuit and the direction of the current depends on the state of the switches.

On the basis of the electrical constraints (voltage and current waveforms) described in the following lines, it can be shown that the characteristics of both semiconductors are those of a thyristor. At the range of the target frequencies (tens of *kHz*) there is no thyristor on the market, so this function is synthesized by combining in series a MOSFET and a diode [11]. The operation of the converter is presented in Figure 2. In the first sequence (Figure 2.a), the thyristor *Th1* is closed, thus the high voltage source supplies the resonant L/C_{eq} circuit (C_{eq} correspond to the capacitances C_{diel} in series with C_{gas}). The DBD current I_{DBD} and the DBD voltage V_{DBD} are controlled by the L/C_{eq} circuit until the gas breakdown. As soon as the discharge is ignited, the gas voltage V_{gas} remains constant at $+V_{th}$, and I_{DBD} and V_{DBD} are then governed by the resonant loop L/C_{diel} (Figure 2.b) until I_{DBD} drops to zero, and the opening of *Th1*. The extinction of the discharge results in the opening of the thyristor *Th1*. Afterwards, there is no evolution of I_{DBD} and V_{DBD} until the closing of *Th2*. When *Th2* is turned *ON*, the DBD and the inductor are short-circuited. The charges accumulated on the dielectric are then used to re-ignite the discharge. The evolution of the electrical waveforms of the DBD is governed by the same resonant circuits: L/C_{eq} as long as the discharge is *OFF* (Figure 2.d), and L/C_{diel} when it is *ON* (Figure 2.e). The ideal waveforms of V_{gas} and I_{dbd} are symmetrical with respect to those obtained in the first half period. An important part of the charges, stored on the dielectrics during (a)-(b), is dissipated in the discharge and the semi-conductors during phase (d)-(e). Therefore in Figure 3 V_{DBD} voltage may not be symmetrical.

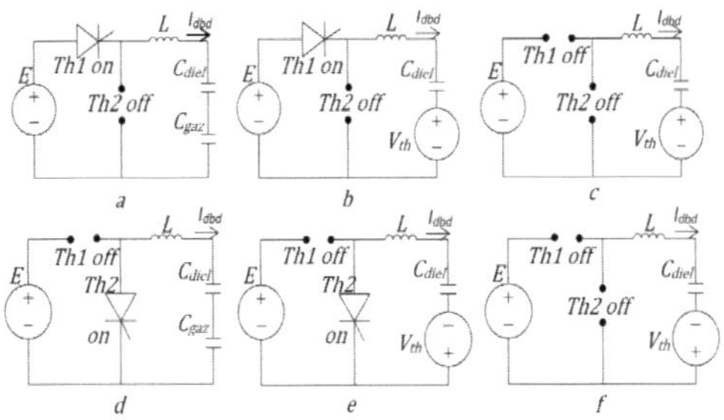

Figure 2: Six-operation stage

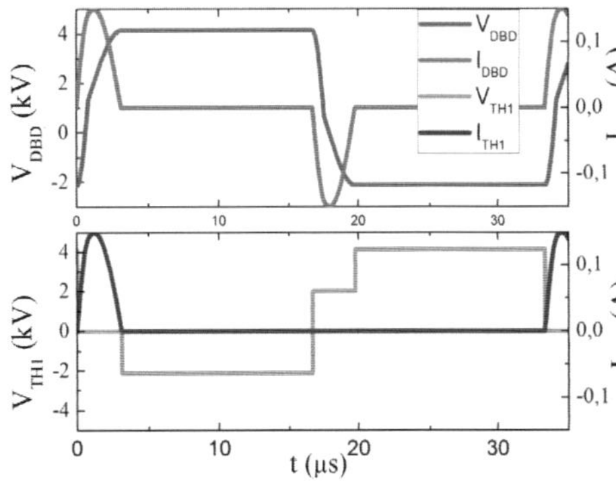

Figure 3: Simulated waveform of the converter

It is important to note that the switch voltage reaches the DBD voltage. In the case of a synthesized thyristor, the MOSFET holds the positive voltage and the diode the negative voltage.

3. Theoretical power dissipated in the discharge

The operation of this converter is studied using the state plan analysis [12] from the model shown in Figure 3. This graphical analysis allows building the trajectories of the current-voltage point of a resonant circuit. Assuming that the steady state is reached, on positive half cycle the operating point describes during a first stage (Figure 4.a) a part of a circle centred on $V_{dbd}=E$ and $I_{dbd}=0$, until the gas breakdown (Figure 4.b); at this time, $V_{gas}=V_{th}$ and the V_{dbd} voltage takes the V_{dbdbr} value.

Once the discharge is ignited, the gas capacitance is charged to $V_{gas}=V_{th}$. During a second stage (Figure 4.c), the part of the circle is centred on $V_{diel}=E-V_{th}$ and $I_{dbd}=I_{dbdbr}$ (Figure 4.d), until $V_{dbd}= V^{+}_{dbd}$ (V_{dbd} voltage when current cancels). In this study, only the positive half-wave is treated. A complete study of the two half periods (work not presented here) has allowed verifying that the current I_{dbd} is symmetrical.

The maximum DBD voltages of the negative and of the positive halves are respectively referred to V^{-}_{dbd} and V^{+}_{dbd}. One can notice that $V^{-}_{dbd}= E- V^{+}_{dbd}$.

Using Figure 4.b and Figure 4.d we can deduce:

$$V^{+}_{dbd} = \frac{4 \cdot V_{th}^{2}\left(1+r\right) - E^{2}}{4 \cdot V_{th} - 2 \cdot E} \qquad (1)$$

With:

$$r = \frac{C_{gas}}{C_{diel}} \tag{2}$$

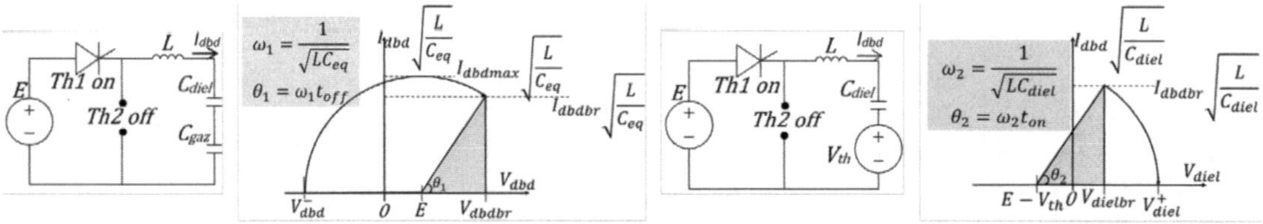

(a) (b) (c) (d)

Figure 4: Phase plan analysis. (a): diagram before breakdown. (b): phase plan before breakdown. (c): diagram during the discharge. (d): phase plan during the discharge

The energy E_n dissipated in the discharge during the first half period is:

$$E_n = V_{th} \cdot \int_{t_{breakdown}}^{t_{off}} i_{dbd}(t) \cdot dt = V_{th} \cdot C_{diel} \cdot \left(V_{diel}\left(t_{off}\right) - V_{diel}\left(t_{breakdown}\right)\right) = V_{th} \cdot C_{diel} \cdot \left(V^+_{diel} - V_{dielbr}\right) \tag{3}$$

with $t_{breakdown}$ the time og the gas ignition and t_{off} the end of the discharge.
To calculate the energy, we need the expression of V^+_{diel} and V_{dielbr}. When the discharge is ignited (between $t_{breakdown}$ *and* t_{off}) the dielectric voltage is:

$$V_{diel}(t) = V_{dbd}(t) - V_{th} \tag{4}$$

so we deduce V^+_{diel}:

$$V^+_{diel} = V_{diel}\left(t_{off}\right) = V^+_{dbd} - V_{th} \tag{5}$$

To calculate V_{dielbr}, we need the expression of V_{dbdbr} obtained from the integration of V_{dbd} between the time of Th_1 switching *ON* (t=0) and the time of the gas ignition (t=$t_{breakdown}$):

$$V_{dbdbr} - V^-_{dbd} = \int_{t=0}^{t_{breakdown}} dV_{dbd} = \frac{1}{C_{eq}} \int_{t=0}^{t_{breakdown}} i_{dbd} \cdot dt = \frac{C_{gas}}{C_{eq}} \cdot \int_{t=0}^{t_{breakdown}} dV_{gas} = 2(r+1) \cdot V_{th} \tag{6}$$

So combining the equation (4) and equation (6) :

$$V_{dielbr} = V^-_{dbd} + V_{th} \cdot (2 \cdot r + 1) \tag{7}$$

From equations (3), (5) and (7), the energy dissipated in the first half period is:

$$E_n = 2 \cdot C_{gas} \cdot V^2_{th} \cdot \frac{E}{2 \cdot V_{th} - E} \tag{8}$$

The energy dissipated in the first half period is equal to the energy dissipated in the second half period so the power P dissipated in the discharge is:

$$P = 4 \cdot C_{gas} \cdot f \cdot V^2_{th} \frac{E}{2 \cdot V_{th} - E} \tag{9}$$

The evolution of the power dissipated in the discharge P *versus* the power supply voltage E is plotted in Figure 5. The circuit presented in Figure 1.b is simulated using PLECS. The simulation confirms the theoretical equation presented in equation (7). It should be noticed that the inductance has no influence on the power. Its value only adjusts the duration of the discharge and the magnitude of the current.

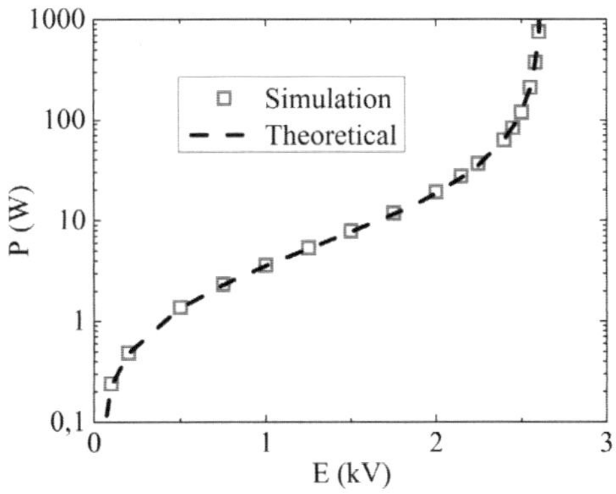

Figure 5: Power transmitted to the discharge versus the power supplies voltage for a frequency of 30 kHz

As long as E is lower than *2000V*, the power delivered to the DBD increases slowly. When E tends to $2V_{th}$, the power tends toward the infinity. In this case at the end of each half cycle the dielectric voltage (or the charge stored on the dielectric) are increased. So, the breakdown appears earlier during the period and the maximum current I_{dbd} of the next discharge increased. Steady state cannot be found.
As shown in equation (9), it is possible to adjust the power P by the frequency and by E. However, it is important to limit E to a value below lower than $2V_{th}$.

4. Experimental validation

A prototype of this power converter has been realized using high voltage switches: SiC power DMOSFET from CREE with a *10kV* maximum drain to source voltage are used. The series connected diodes C4D02120A, holding up to *1700V* from CREE are used. Throughout the whole paper (for experiment or for simulation), the inductance value is *17mH*. At *30kHz* and with the lamp used here, the duration of the pulse is always lower than the half period. Therefore, the switch operates really as a thyristor. The volume and the weight of the power supply are very weak compared to a power supply with a transformer.
The voltages are measured using high-voltage differential probes, TESTEC TT-SI 010 A, and the current is measured using a 4100 Pearson probe. Electrical waveforms are recorded using a LECROY Wavesurfer 24XS oscilloscope. The constant voltage source E is an adjustable power supply (SFRDO-F-2000, Technix brand).
The electrical waveforms (V_{dbd}, I_{dbd} and V_{th1}), measured at *30kHz* and *E=3 kV*, are presented in Figure 6. The time evolution of V_{th2} is the same as for V_{th1}, delayed by a half period. When the discharge turns *ON* the waveforms are very close to those presented above. According to the previous theoretical analysis, it has been unexpected to observe some oscillations in all the waveforms when the discharge and the switches turn *OFF*. This is because the diagram of the power supply presented in Figure 1.b is simplistic and does not take into account the parasitic elements. As can be seen, the oscillations cause over-voltage across the semiconductors. This can be a serious issue for transformerless power supply, because high voltages switches must sustain this over-voltage and are currently limited to *10kV*.
In Figure 7 we present a simulation model of the power supply taking into account all the parasitic capacitances. The value of the MOSFET capacitance C_{ossi} and of the diode capacitance C_{di} are taken from the datasheet and arranged in a table indexed by semiconductor voltage. The parasitic capacitance of the inductance C_{para_L} is measured by an impedance bridge and is equal to *5pF*. The

probe capacitance C_{probe} is also measured using two current probes (one upstream and one downstream the voltage probe). Finally a parasitic capacitance C_{para} is inserted between the probe and the DBD. This last parasitic capacitance takes into account the high voltage connection to the DBD (wire, high voltage terminals ...). C_{para} is measured when the discharge is *OFF* by a basic current voltage method and its value is *15pF*.

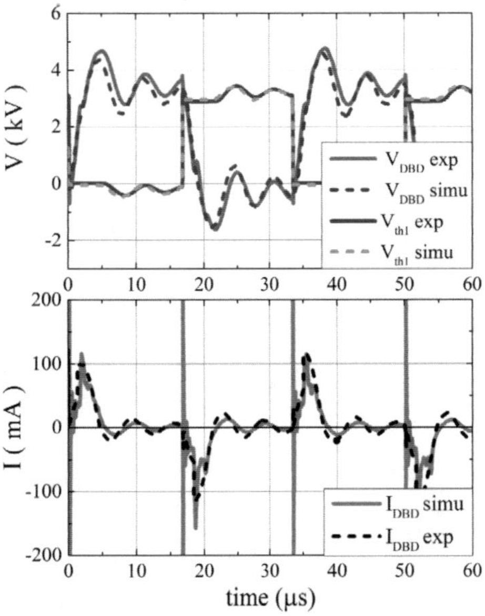

Figure 6: Electrical waveform: comparison between experimental measurement and simulation of the diagram of **Figure 7**.

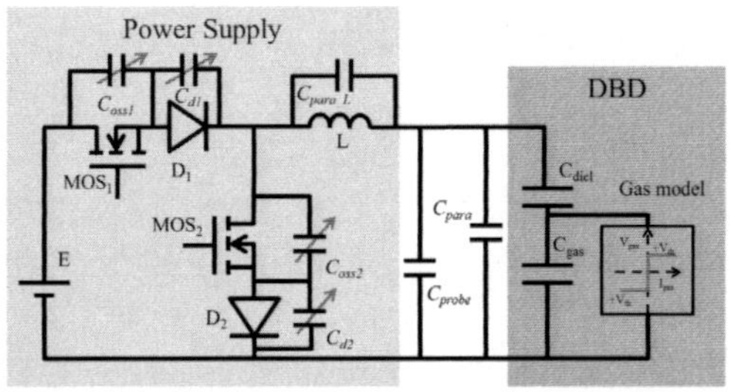

Figure 7: diagram of the power supply tacking into account al the parasitic capacitance.

The power supply with all the parasitic capacitance presented in Figure 7 is simulated with PLECS. As presented in Figure 6, this simulation accurately fits the measured waveforms: same amplitude, same oscillation frequency. The oscillations observed after the extinction of the discharge are therefore due to the parasitic capacitance presented in Figure 7. If all these parasitic capacitances are not taken into account, the frequency and the amplitude of the oscillations are not accurate. Nevertheless, in order to simplify the analysis, we consider in the following sections separately the role of each parasitic capacitance.

5. Influence of the parasitic capacitance of the switches

In this part we use the basic thyristor model developed by PLECS. For each thyristor, Th_1 and Th_2, only a constant capacitance C_{p_Th} takes into account the parasitic effects as seen in Figure 8. In Figure 10.b, the simulated waveforms of V_{DBD}, V_{diel}, V_{Th2} and, i_{Th1} are presented for the first half period (when Th_2 is in *OFF* state) and for a frequency of *30kHz*. When Th_1 is turned *ON*, the waveform does not change with or without parasitic capacitances. During this phase, the parasitic capacitance of Th_1 is short-circuited and the one of Th_2 is charged to *E*: no charges are transferred between these parasitic capacitances and other elements of the circuit. As long as Th_1 is *ON*, its parasitic capacitance plays no role. When the current of Th_1 drops to zero, the discharge extinguishes and Th_1 blocks spontaneously. So all switches are *OFF* and electric charges move between the DBD and the switches parasitic capacitances. Indeed, voltage is not imposed on C_{p_Th}. The charge transfer leads to oscillations that are visible in all the electrical waveform. Therefore the parasitic capacitances of the switches are the only responsible of these oscillations. It is not shown here, but it is easy to understand that the same phenomenon appears on the second half period (Th_1 remains *OFF*).

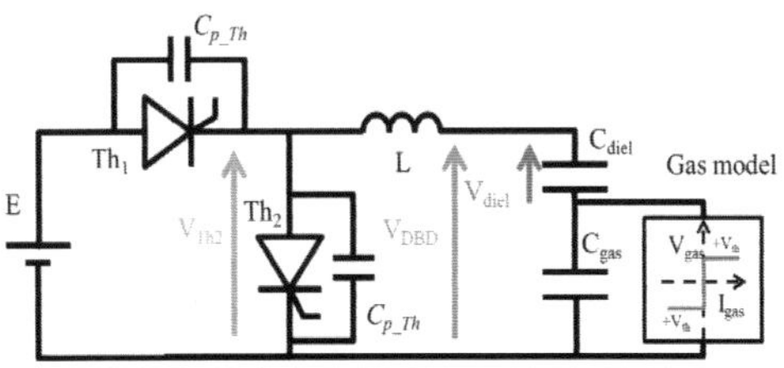

Figure 8: Power supply with only the parasitic capacitances of the switches.

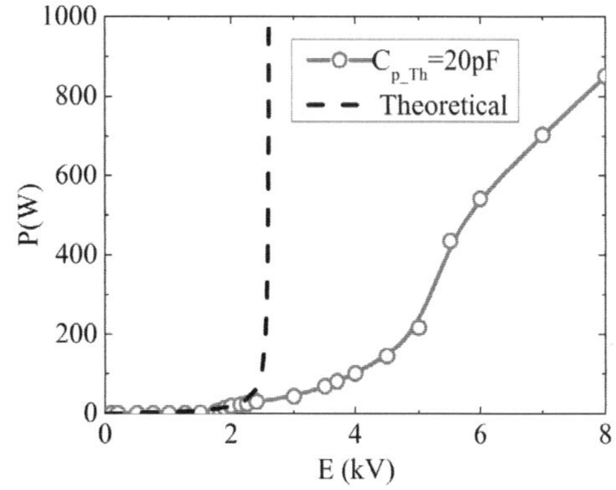

Figure 9: Power transferred to the discharge. Full line: with a parasitic capacitance of 20 pF. Dashed line: Without parasitic capacitances (equation (9))

In Figure 9, we present the evolution of the power *versus* the constant voltage E for a parasitic capacitance C_{p_Th} of *20pF* (red curve). This value corresponds to the order of magnitude of a high voltage switch.

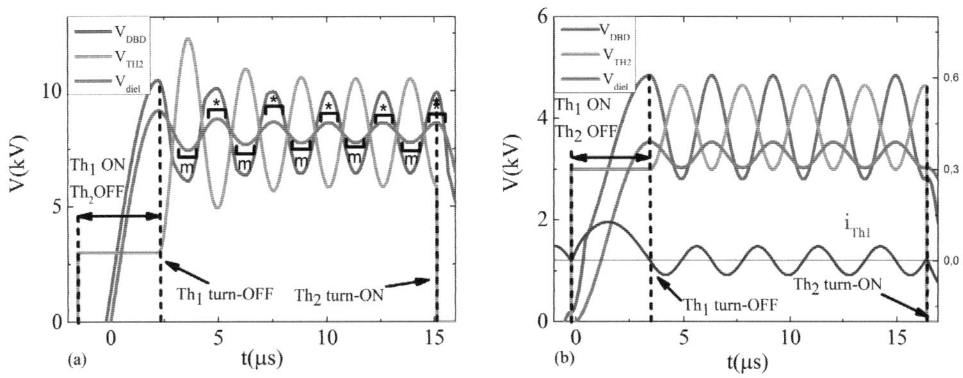

Figure 10: Simulated electrical waveform of the power supply with a parasitic capacitance. First half period when Th_1 turned OFF and Th_2 turned ON. Switching time are indicated (a): $C_{p\_Th}=7.8pF$. Bracket *: local maxima of V_{diel}. Bracket m: local minima of V_{diel}. (b): $C_{p\_Th}=13pF$.

As can be seen in Figure 9, parasitic capacitances of the switches clearly affect the power transferred to the discharge. With switch parasitic capacitances, the power is lower than without switch parasitic capacitances. Moreover with switches parasitic capacitance, the powers do not present a hyperbolic trend when E is close to $2V_{th}$. The power supply can operate in a wider range of voltage E. Indeed, as soon as all the thyristors are open, the parasitic capacitances of the switches discharge the dielectrics of the DBD. In any case, compared to the theoretical case, fewer charges are stored on the dielectrics when a thyristor is closed. If fewer charges are available to ignite the next discharge, the energy dissipated will be lower. This phenomenon explains why the power dissipated in the discharge with C_{p_Th} is lower than the power dissipated in the discharge without parasitic capacitances.

Figure 11 presents the evolution of the power transferred to the discharge versus C_{p_Th} for a frequency of *30kHz* and *E=3kV*. Overall, the power tends to decrease with C_{p_Th} but there are local maxima where the power strongly increases.
To understand these power peaks we have to look carefully at the temporal evolution of V_{Th2} and V_{diel} presented in Figure 10.a and Figure 10.b. The Figure 10.a corresponds to a power peak, and Figure 10.b corresponds to a power deep. As can be seen, the numbers of local minima and maxima on Figure 10.a and Figure 10.b are different. The parasitic capacitance changes the oscillations frequency and the number of oscillating periods during the *OFF* time of the discharge (when *Th1* and *Th2* are in *OFF* state). In some cases, Th_2 is closed when V_{diel} is maximal and there is a power peak (Figure 10.a), in other cases Th_2 is closed when V_{diel} is minimal and there is a power deep (Figure 10.b). Therefore we observe a power deep when fewer charges are stored on the dielectric, and at the opposite we observe a power peak when more charges are present on the dielectric. This further regard confirms that the power dissipated in the discharge strongly depends on the charges stored on the dielectric before the closing of the switches.

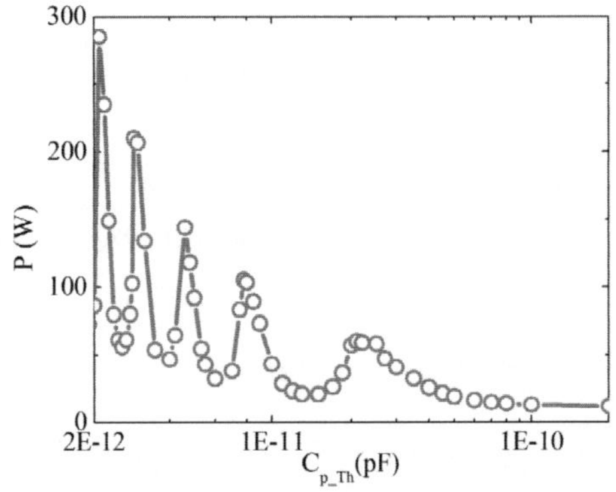

Figure 11: Evolution of the transferred power versus the switch parasitic capacitance for E=3kV and f=30kHz.

The parasitic capacitances of the switches modify the power transferred in the discharge. In some case the values of the parasitic capacitances amplify the power transferred in the discharge. Improving the power transfer requires to take these parasitic capacitances from the very early stage of the design. Anyway, it is impossible and illusory to imagine adjusting perfectly the value of the parasitic capacitances.

6. Influence of the voltage probe capacitance

The probe capacitance C_{probe} or the parasitic capacitance C_{para} (which consists of the connection and the wire) do not substantially change the situation explained above; their impacts is mainly a change of the resonant frequency that defines the duration of the current pulses.
If the switches parasitic capacitances are not considered, there are no additional oscillations (with respect to the theoretical waveforms of Figure 3) when all the switches are in *OFF* state; the *P(E)* plot (power *P* versus *E* voltage) keeps the hyperbolic shape presented in Figure 5 and shown again on Figure 9.
With switches parasitic capacitances, the oscillations always exist when both switches are *OFF*.
C_{probe} affects the oscillation frequency and the C_{para} or C_{probe} capacitances shift lightly the curve presented in Figure 11, but do not change fundamentally the operation of the power supply.
Moreover, if a voltage probe is connected to switch terminals, its parasitic capacitance will play the same role as the parasitic capacitance of the switch.

7. Influence of the parasitic capacitance of the inductance

Depending on the presence of the switches capacitances, the parasitic capacitance of the inductance affect markedly the operating of the power supply.

7.1. Without switches parasitic capacitances

We use PLECS to simulate the power supply presented in Figure 1.b but we add a capacitance of $C_{para\_L}=5pF$ in parallel with the inductance. The time evolution of V_{Th2}, V_{DBD}, V_{diel} and I_{dbd} is presented in Figure 13. The time evolution of V_{th1} is the same that the time evolution of V_{th2} delayed by a half period.

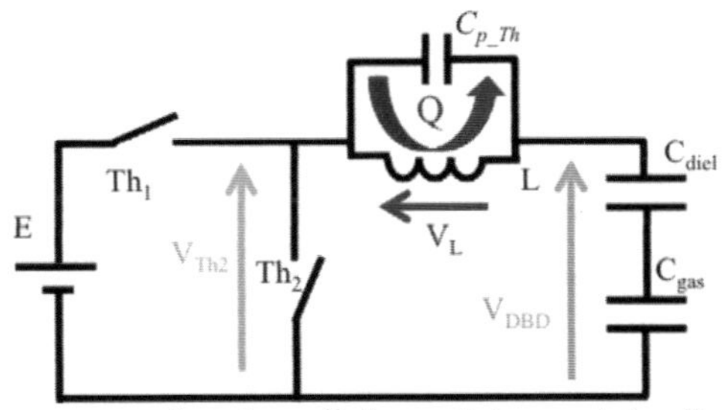

Figure 12: Charges transfer when all the switches and the discharge are OFF

When all the switches are *OFF* (Figure 12), the oscillating circuit consists on L in parallel with only C_{para_L}. The moving charges Q are presented in Figure 12. Because all the switches are in *OFF* state, no current flows through the DBD and the DBD voltage remains constant. The switches voltage is equal to the DBD voltage plus the voltage of oscillating circuit L/C_{para_L} (Figure 13).

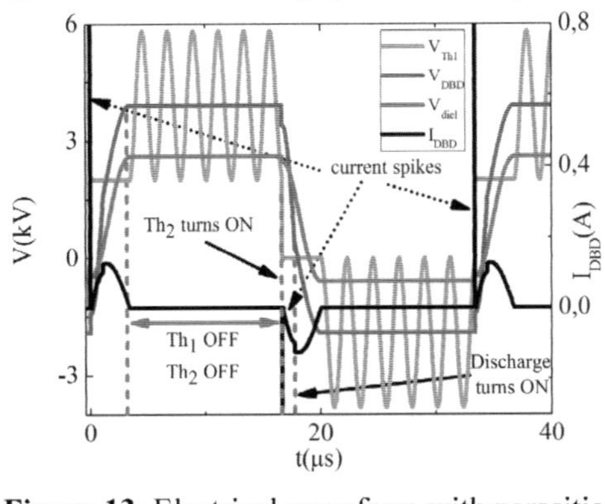

Figure 13: Electrical waveform with parasitic capacitance of the inductance. E=2kV, f=30 kHz.

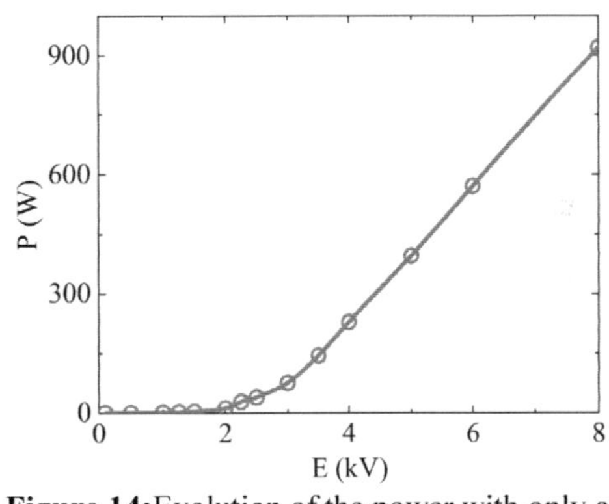

Figure 14: Evolution of the power with only an inductance parasitic capacitance C_{para_L}=5pF. f=30kHz

When one switch is turned *ON* we observe a current spike on I_{DBD}. This is due to a sudden change of the voltage applied on the inductance parasitic capacitance. According to the current ratings of the semiconductor, these spikes could damage the switches. Moreover, when the switching appears, a strong and swift voltage step is directly applied on the series circuit consisting in the DBD and the inductance. Without C_{para_L} the inductance L sustains the totality of the voltage step. With C_{para_L}, this voltage step is applied to the 3 capacities in series: C_{para_L}, C_{gas}, C_{diel}. As we can see in Figure 13, it causes a slight drop of V_{DBD} and V_{diel} as soon as Th_1 or Th_2 turns *ON*. As explain above, a drop of the dielectric voltage causes a limitation of the power. As shown in Figure 14, with only the parasitic capacitance of the inductance, the power plot *P(E)* are not an hyperbolic trend; C_{para_L} limits the power transferred to the discharge.

7.2. With switches parasitic capacitances

The operating is nearly the same as the one described in the part 5 (with only the parasitic capacitances of the switches). However, the current spike is now present as explained above in 7.1. Due to its low value, C_{para_L} affect scarcely the operation of the power supply on the power transfer.

8. Conclusion

Aimed at limiting the negative effects of the high voltage step up transformer, an innovative transformerless topology for DBD has been proposed.

This specific resonant inverter operates as an instantaneous current source and supplies the DBD with positive and negative current pulses. The properties of these pulses (magnitude, duration, frequency f, direction) are fully adjustable using the design (values of the circuit's components) and operating degrees of freedom (firing orders, magnitude E of the DC source).

10kV high voltage SiC MOSFETs and diodes are used to synthesize the thyristor-like characteristic of the power switches. These devices, as well as the resonant inductor, introduce parasitic capacitances. The most important points concerning these parasitic capacitances are:

- An increase of the voltage stresses on the semiconductor SiC devices;
- A modification of the output power due to the drop of the charges stored on the dielectrics of the DBD. This charges are diverted to the parasitic capacitances of the switches when the latter are *OFF*;
- A current spikes in the switches at turn *ON*, due to the sudden voltage step of the inductance's parasitic capacitance.

As a matter of fact, all these side effects substantially modify the operating conditions of the semiconductors and need to be early accounted at design step. A special care at implementation stage needs to be taken in order to reduce the parasitic capacitances as much as possible.

References

[1] Kogelschatz U 2003 *Plasma Chem. and Plasma Proc.* **23** 26

[2] Williamson J, Trump D D, Bletzinger P and Ganguly B N 2006 *J. Phys. D: Appl. Phys.* **39** 4400

[3] Lomaev M I, Skakun V S, Sosnin E A, Tarasenko V F, Shitts D V and M V Erofeev 2003 *Phys. Usp.* **46** 193

[4] Massines F, Sarra-Bournet C, Fanelli F, Naudé N and Gherardi N 2012 *Plasma Process. and Poly* **9** 1041

[5] Corke C T, Post M L and Orlov D M 2008 *Exp. Fluids* **46** 1

[6] Bonnin X, Piquet H, Diez R, and Florez D 2013 *15th European Conference on Power Electronics and Applications* (Lille) pp. 1-9

[7] El-Deib A, Dawson F and Zissis G 2011 *IEEE Energy Conversion Congress and Exposition* (Phoenix) pp 1124-1131

[8] Díez R, Salanne J P, Piquet H, Bhosle S, and Zissis G 2007 *Eur. Phys. J. - Appl. Phys.* **37** 307

[9] Bonnin X, Piquet H, Naudé N, Bouzidi M C, Gherardi N and Blaquière J M 2013 *Eur. Phys. J. Appl. Phys.* **64** 10901

[10] Piquet H, Bhosle S, Diez R, and Erofeev M V 2010 *IEEE Trans. Plasma Sci.* **38** 2531

[11] Cousineau M, Diez R, Piquet H, Durrieu O 2012 *IEEE Trans. Ind. Electron.* **59** 1920

[12] Chéron Y 1992 *Soft Commutation* (Springer Science & Business Media)

First Breakthrough for Future Air-Breathing Magneto-Plasma Propulsion Systems

B Göksel[1*], I Ch Mashek[2]
[1] Electrofluidsystems Ingenieurbüro Göksel, Berlin, Germany
[2] St. Petersburg State University, St. Petersburg, Russia

[*]Email: berkant.goeksel@electrofluidsystems.com

Abstract. A new breakthrough in jet propulsion technology since the invention of the jet engine is achieved. The first critical tests for future air-breathing magneto-plasma propulsion systems have been successfully completed. In this regard, it is also the first time that a pinching dense plasma focus discharge could be ignited at one atmosphere and driven in pulse mode using very fast, nanosecond electrostatic excitations to induce self-organized plasma channels for ignition of the propulsive main discharge. Depending on the capacitor voltage (200-600 V) the energy input at one atmosphere varies from 52-320 J/pulse corresponding to impulse bits from 1.2-8.0 mNs. Such a new pulsed plasma propulsion system driven with one thousand pulses per second would already have thrust-to-area ratios (50-150 kN/m²) of modern jet engines. An array of thrusters could enable future aircrafts and airships to start from ground and reach altitudes up to 50km and beyond. The needed high power could be provided by future compact plasma fusion reactors already in development by aerospace companies. The magneto-plasma compressor itself was originally developed by Russian scientists as plasma fusion device and was later miniaturized for supersonic flow control applications. So the first breakthrough is based on a spin-off plasma fusion technology.

Introduction

Magneto-plasma compressors (MPC) were originally developed as quasistationary plasma accelerators for fusion applications [1, 4]. They consist of coaxial electrodes representing an electromagnetic Laval nozzle analogue [4, 8], see Figure 1. In the air regime with initial pressures from 0.1-10 Torr the plasma jet has a lifetime of 50-100 μs, can reach velocities from 5-20 km/s and high densities from 10^{17}-10^{18} cm^{-3} [8]. In the plasma focus the pinching discharge can be compressed to pressures of about 100-150 bar [8]. High speed plasma jets generated by magneto-plasma compressors (MPC) can be applied for plasma fusion devices [1-6, 15, 26-28], high-enthalpy plasma-material interactions, surface modification [7, 15, 25, 27], supersonic flow control [8-11] and air-breathing propulsion [12-14].

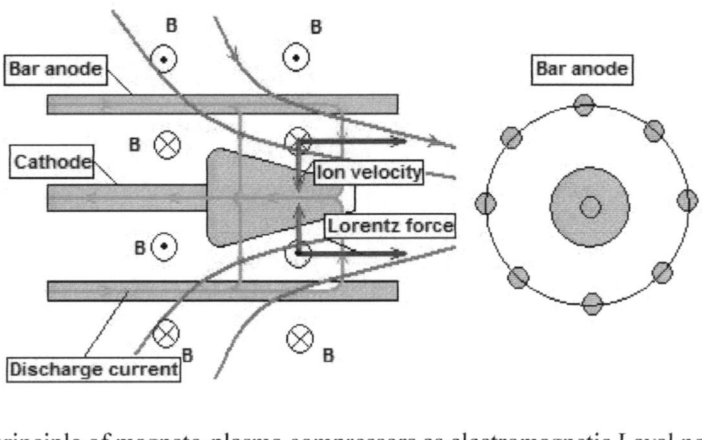

Figure 1. Physical principle of magneto-plasma compressors as electromagnetic Laval nozzle analogue [8].

Motivation

Almost all known MPC types operate under high vacuum conditions (10^{-5}-10 mbar) with very high discharge currents (10-200 kA) and low repetition frequencies (0,01-0,1 Hz). The overall system in all cases is based on bulky high-current switchers, heavy high voltage capacitor banks with tens of kV, complex control and diagnostics devices [1-7]. Nevertheless, all the previous work gave the main inspiration to develop new lightweight, highly miniaturized, lower voltage powered stable MPC-based plasma jet generator systems for flow control and propulsion purposes. The main initiation was given in 2010 by an informal request from an European aerospace company for a stratospheric high-thrust plasma propulsion with tens to hundreds of Newton. At the time this technology was not available and the request could not be addressed but the idea was born to develop a new propulsion system which combines electrohydrodynamic barrier discharge and field electron emission effects with magnetohydrodynamic flux compression aspects. The first breakthrough milestone goal was to realize a MPC system with 5-10 Hz repetition frequency which is able to operate under atmospheric pressures from 0.1-1.0 bar [13]. So far there was no known work about MPC operating at high-pressure and high-frequency. This challenge and the industrial request for a high-altitude propulsion systems was the main motivation to start a R&D programme for pulsed plasma "detonation" thrusters.

It is not widely known that a hydromagnetic Rankine-Hugeniot model for detonation and deflagration can be also used to describe gas-fed (air-breathing) coaxial plasma accelerators which are based on pinching dense plasma focus discharge devices like the magneto-plasma compressor. The gas is in this case is resistively heated by the pinching discharge instead of the burning in the combustor. The hydromagnetic shock caused by the high magnetic pressure ionizes and compresses the gas volume ahead of itself [19-22]. Nevertheless, all the known studies focus on high-vacuum conditions with plasma jet velocities exceeding 100-200 km/s.

Recent studies about sub-millimeter dense plasma focus (DPF) devices also reveal attempts and proposals towards higher pressures of 10-1000 Torr in hydrogen though limitations in confinement due to increased collisions at high pressures are expected [23]. Nevertheless, the same authors could recently demonstrate a DPF operating in 50-190 Torr Helium [24]. This specific type of pinching plasma accelerator is producing short nanosecond discharge pulses which are also relevant for x-ray and short pulse neutron emissions. DPF devices and their applications are again reconsidered for investigation in large-scale plasma fusion and space propulsion programs [26-31]. This is a further motivation to study air-breathing magneto-plasma jet thrusters.

In previous experiments the authors used microsecond discharges for internal initiation of the propulsive main discharge and could already increase the MPC operation pressure from 30 to 250 Torr (0.33 bar), see Figure 2a-2c [13]. The use of nanosecond discharge was the next logical step which was also motivated by the fact that in experiments to chemical pulse detonation engines distributed or transient nanosecond spark discharges are significantly more efficient for detonation initiation than localized microsecond spark discharges of comparable pulse energy [17]. So high repetition nanosecond pulsed discharge are already in the focus of groups working on efficient combustion processes for several years [18], though no working group has obviously ever combined the nanosecond know-how with pinching dense plasma focus discharges.

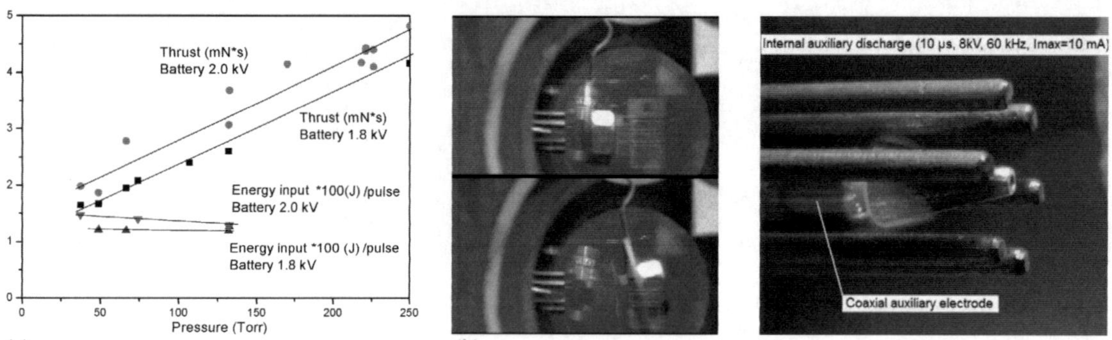

(a) (b) (c)

Figure 2a-2c. Thrust and pulse energy deposition in dependence of the air pressure for different battery voltages with internal microsecond sliding discharge exitation [13].

IOP Conf. Series: Journal of Physics: Conf. Series **825** (2017) 012005 doi:10.1088/1742-6596/825/1/012005

General experimental setup

The new MPC has an outlet diameter of 14 mm and contains six coaxial bar anodes (each from 3mm Copper rods) and one conical Copper cathode with a maximum diameter of 7 mm. The minimum distance between each anode and cathode is 2 mm. Following the electrode size and distances we can name this configuration (3-2-7) mm. The overall length of the new compressor is 80 mm, see Figure 3a. The MPC location inside an acrylic vacuum chamber is shown in Figure 3b and 3c.

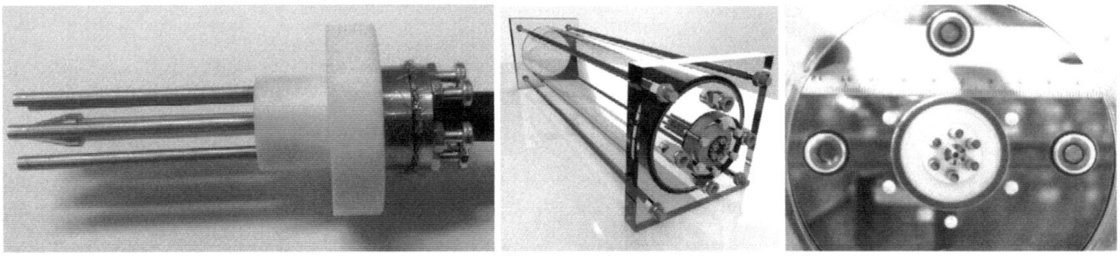

(a) (b) (c)

Figure 3a-3c. Tested MPC and test vacuum chamber.

The MPC is designed to work under pressures from 0.1-1.0 bar. The related self-breakdown threshold for the new (3-2-7) configuration is below 5kV. A nanosecond (ns) high voltage pulse generator (NPG-18/3500N, Megaimpulse Ltd, Russia) is used to induce a homogeneous, transient discharge for internal MPC excitation. High-frequency nanosecond pulse discharges have the unique property to induce transient self-organization of plasma channels between the electrodes [16, 17], see Figure 4a-4c. The transient ns-discharge is driven by -15 kV, 3.5 kHz pulses with 4 ns rise time and 10-20 ns pulse length for all test pressures. The full dynamic behavior of the ns-discharge will be addressed in a separate paper. The general scheme of experimental setup for the new MPC investigation is presented below in Figure 5.

(a) (b) (c)

Figure 4a-4c. Photos of transient atmospheric nanosecond pulse discharges for internal MPC excitation.

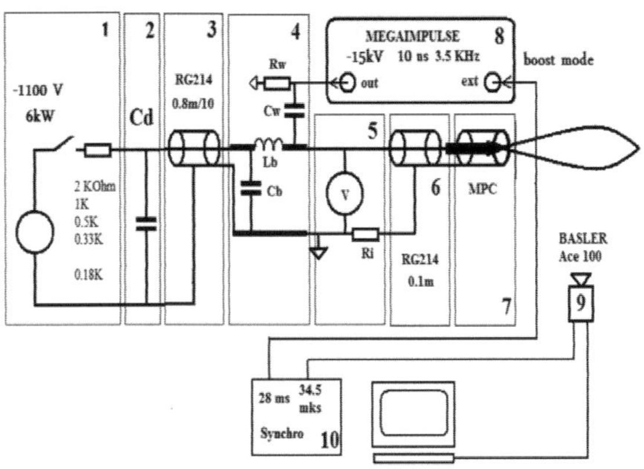

Figure 5. General scheme of the setup.

The new MPC (7) is connected to a cable collector (4) through a short piece of coaxial cable (6) (0.1 m RG214U) and a discharge current/voltage measuring circuit (5) (current resistor 0.75 mOhm, voltage sensor PMK-14KVAC). The cable collector contains a high-frequency blocking filter (L_b=150 nH, C_b=1,1 nF), a separating capacitor C_w (220 pF, 16 kV), a pull-down resistor R_w (75 Ohm) and a low-inductance Copper plate which leads the current for the capacitor bank (battery) output cables. The cable collector (4) itself is connected with the main discharge battery by ten coaxial cables (3) having each 0.8 m length (RG214U). The main discharge capacitor bank consists of 10 impulse capacitors with 2700 μF and 1200V maximum voltage. This capacitor bank (2) is charged through the resistor R_c and a voltage regulator with a maximum output voltage of 1100 V and 6 kW power.

The nanosecond pulse generator (8) (NPG-18/3500N) operates in the boost regime (external trigger). The external synchronization input (28 ms) generates about 98 high voltage (HV) ns-pulses which again induce the homogeneous, transient plasma channels between the cathode and coaxial anodes. The blocking filter in the cable collector (4) saves the main battery from damage through each of the nanosecond HV pulses.

The capacitor bank (2) has a total capacity of 2700 μF, an internal resistance of 0.3 mOhm and an internal inductance of 8.5 nH (10 capacitors in parallel connection). The cables connecting the battery with the cable collector have a self-inductance of 20 nH and a resistance of 0.6 mOhm (10 RG214U cable with each 0.8 m in parallel). The cable collector has the resistance of 0.02 mOhm and a self-inductance of 6.6 nH. The inductance of the blocking filter is 150 nH. The short connecting cable for the MPC (0.1 m RG212U) has 25 nH and 0.7 mOhm.

Visualization setup and results

Visualization of MPC induced plasma jets was performed with a Basler Ace100 camera (9 in Figure 5) having a frame exposition able to synchronize with the nanosecond HV pulse generator working in a boost regime with 28 ms. The camera itself was launched with a synchronizing circuit (10 in Figure 5) using short TTL pulses with a duration of 34.5 μs. The end of a pulse launches the boost regime of the ns pulse generator (8 in Figure 5). To minimize EM interferences from the nanosecond pulse discharges to the camera and computer electronics the synchronization pulse was transferred to the nanosecond HV pulse generator through an optical fiber coupling.

In Figure 6a-6c, a general view of the high pressure plasma jet is presented. All pictures were taken with a dense red glass filter KS-15 and minimal camera lens diaphragm.

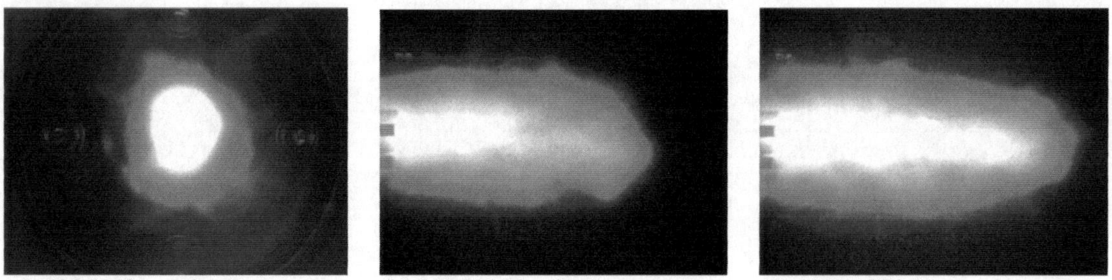

(a) 400 V (front view) (b) 400 V (side view) (c) 500V, 25 kA

Figure 6a-6c. View of MPC plasma jets for different battery voltages U_b at an air pressure of 450 mbar.

Usually, the main discharge is arising after 2-10 pulses of the internal ns-excitation. At one atmosphere the MPC plasma jets have a similar character but higher intensity than under a lower pressure of 450 mbar, see Figure 7a-7c. In Figure 7a, it can be clearly seen that a pinching plasma jet with a compression focus area already exists for a battery voltage as low as 300 V. The pinching plasma can be even more clearly seen through IR and UV filters, see Figure 8a-8b. Figure 9a-9b show a visualization of the nanosecond excitation (Figure 4a) and main discharge through a blue filter. In all previous studies the battery voltage was much higher in the range of at least 1.8 kV [13]. Nevertheless, the capacitor bank can provide voltages up to 1.2 kV.

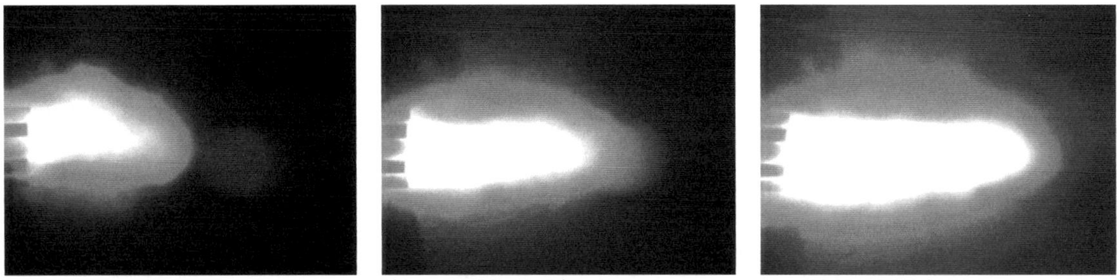

(a) 300 V, 11.0 kA, (b) 400 V, 18.5 kA, (c) 500V, 25.0 kA.

Figure 7a-7c. Plasma jets for different battery voltages U_b at 1 bar atmospheric pressure through a red filter.

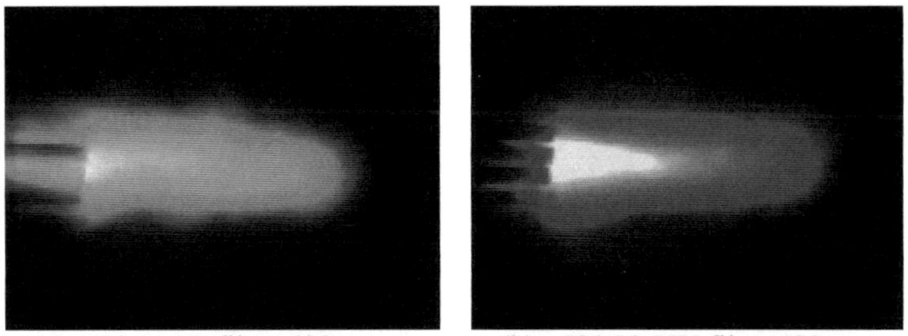

(a) 500 V, IR filter 680 nm, (b) 500 V, UV filter 480 nm.

Figure 8a-8b. Plasma jet for U_b=500 V at 1 bar atmospheric pressure through IR and UV filters.

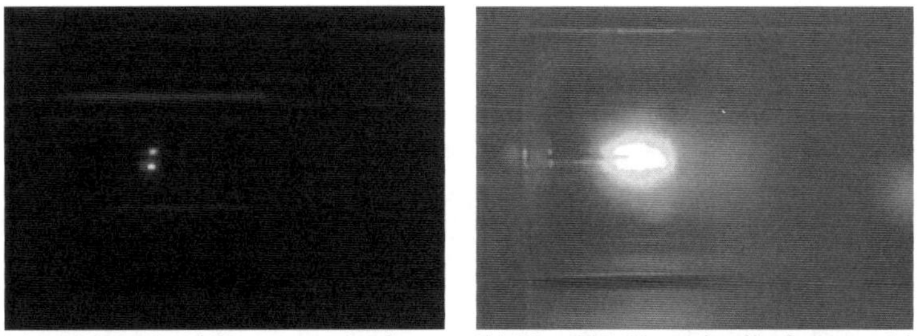

(a) -15 kV, 3.5 kHz ns-excitation, (b) U_b=300 V, main discharge with I_d=11.0 kA.

Figure 9a-9b. Excitation and ignition of main discharge at 1 bar atmospheric pressure through a blue filter.

Current and voltage measurements and simulations

The discharge current for the MPC has a weak dependence from pressure, and mainly depends from the discharge voltage. A typical presentation of the discharge parameters are shown in Figure 10a.

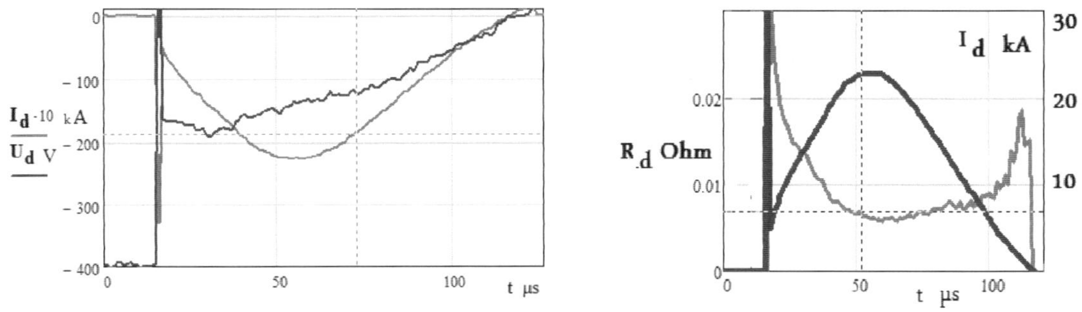

(a) (b)

Figure 10a and 10b. MPC discharge for U_d=400 V (battery with 2700 µF) at 1 bar atmospheric pressure.

In Figure 10b, the nonlinear resistance (red line) of the pinching high-current arc discharge is shown over time. The discharge resistance R_d dramatically decreases when the discharge current reaches the higher peak levels of 15-18kA in this specific case, Figure 10b. This effect is related to the tangential component of the strong magnetic field arising between MPC electrodes. The intrinsic tangential component of the magnetic flux density is the most important factor in this type of pinching plasma accelerators.

For a first estimation the value of the magnetic fields at different discharge currents were simulated in COMSOL, Figure 11a-11b. The electrodynamic model of the MPC contains six circuits with anode rods and a common center cathode. The plasma parameters were estimated from experimental data of current voltage characteristics. The average cross-section for each of the six plasma channels is about 4 mm^2 with a resistance of 60 mOhm. The current for each circuit is equal to 1/6 of the overall discharge current (25-30 kA).

The experimentally obtained relationship between the nonlinear discharge resistance and the discharge current clearly shows that the generation of a flux compression plasma jet in the MPC is only possible at a certain level of the tangential magnetic flux density in the discharge gap. Minimum 0.7-1.2 T is necessary for the investigated plasma accelerator. The distribution of the tangential component of the magnetic flux density in the electrode gap for a discharge current of 30 kA is shown in Figure 11a for a cross-section placed 1 mm downstream of the plasma channels. Figure 11b shows the radial distribution of the tangential component of the magnetic field for different discharge currents (direction cathode center to anode center).

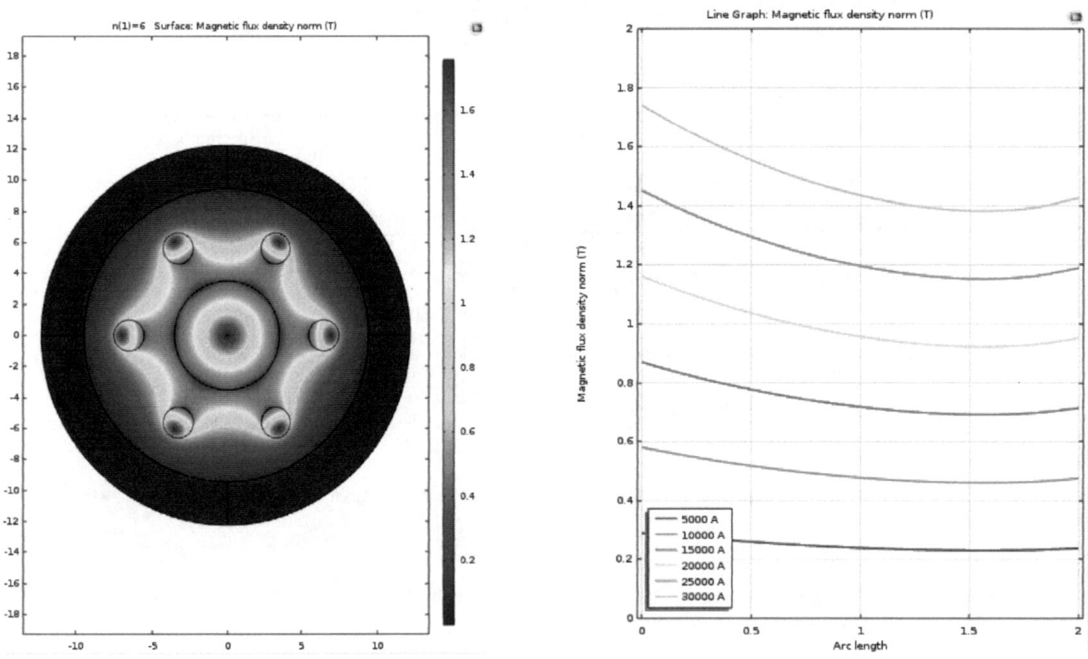

(a) (b)

Figure 11a and 11b. The distribution of tangential magnetic fields between the MPC electrode gap.

The results of simulations and experiments show that there are strict requirements on the geometry of the discharge electrode. A flux compression with a low energy loss can be reached for discharge currents higher than 15-18 kA. This level corresponds to a minimum tangential magnetic field of 0.7-0.9 T. Particularly, the discharge voltage in the working gap at the proper geometry becomes so small that it causes low energy losses. Thus, the energy efficiency (ratio between discharge energy and energy of the capacitor bank) for the (3-2-7) mm MPC version with six-anodes at low voltage is about 0.85-0.90 but is decreasing for higher voltages.

IOP Conf. Series: Journal of Physics: Conf. Series **825** (2017) 012005 doi:10.1088/1742-6596/825/1/012005

The capacitor bank voltage U_b here was varied from 200 to 600 V, though the maximum possible value is 1200 V. The overall energy efficiency for other voltages can be estimated or calculated using the formula (1), the W_d, U_d values from Figure 14 and $C_b = 2700\mu F$.

$$\eta = W_d / W_b = \int I_d(t) \cdot U_d(t)\, dt / [C_b \cdot U_b \cdot U_b / 2] \qquad (1)$$

Impulse bit measurements

The impulse bit of the new MPC was investigated using a free pendulum with 55 mm length and a mass of 15 g (see Figure 12a-12b). The pendulum deflection angle varied from 5° to 25°. Such small angles allow the use of the simplest model for a mathematical pendulum. The method used for thrust measurement is comparatively simple but has several sources for systematic errors. The first is the kind of plasma jet-surface interaction (elastic, non-elastic). An absolutely inelastic interaction of the plasma jet with the pendulum surface was assumed. The second connects with the uncertainty of contact point for the jet-surface interaction because the direction of the generated plasma jet shows some fluctuations. And the third main reason for errors is the stochastic nature of the impulse arc discharge at high pressures of 0.2-1.0 bar. The estimated measurement accuracy is 10-15 %. Figure 12c shows a series of small motion pictures taken from a video for a variant with a more lightweight pendulum. Conductive and insulating pendula made the same impulse bit. Figure 13 shows the measured impulse bit for different battery voltages and different atmospheric pressures. The first results show a comparatively weak dependence between impulse bit and atmospheric pressure in the measured range from 175-1000 mbar.

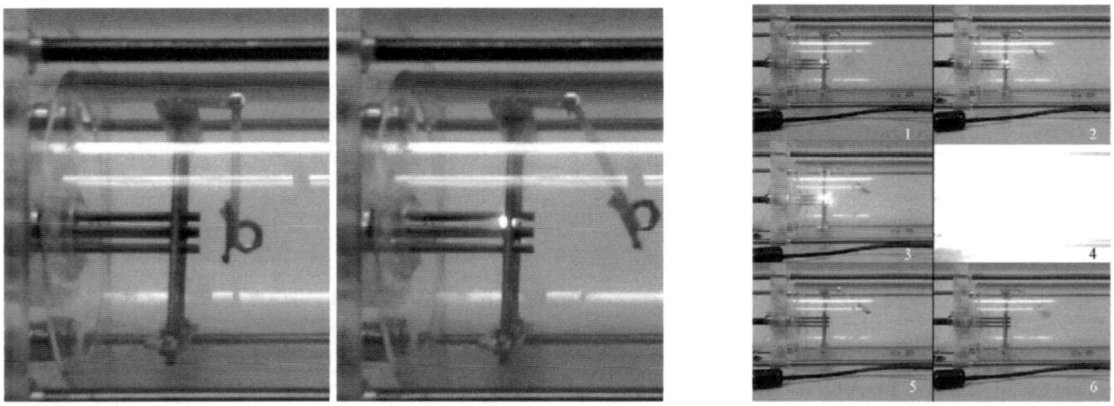

(a) 0 V, 1000 mbar (b) 400 V, 1000 mbar (c) 400 V, 1000 mbar

Figure 12a-12c. Measurement of impulse bit (Ibit) and thrust demonstration using a pendulum.

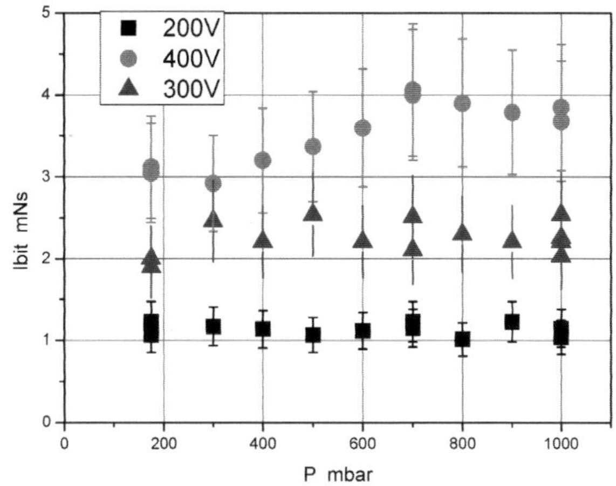

Figure 13. Measured impulse bit for different battery voltages and different atmospheric pressures.

Summary and outlook

Figure 14 gives a summary of the main results for the maximum discharge current $I_{d,max}$, the impulse bit and the discharge energy W_d. The input energy W_b (proportional to the square of the battery voltage U_b) is dramatically increasing for higher input voltages, see also formula (1). The measured discharge energy W_d is also increasing in a similar way. But at higher voltages beyond 400 V the energy loss is rising too. So the effective energy available to initiate the discharge is not proportional to the input energy. The kink in the discharge energy slope at 400 V is caused by the higher energy losses at 500 and 600 V. In these cases, the energy efficiency is only 0.74 and 0.66 compared to the much higher efficiencies at voltages below 500 V.

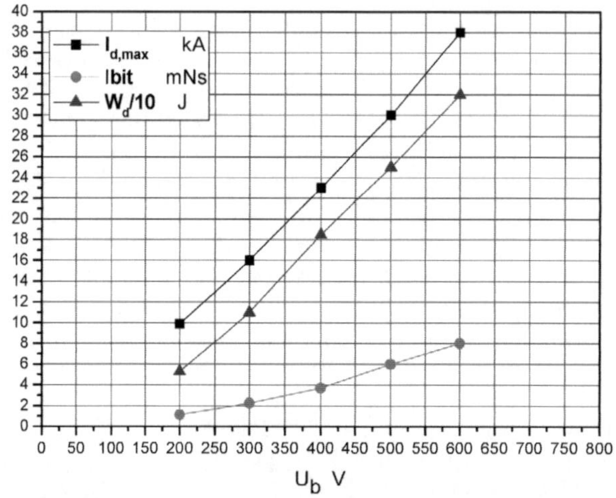

Figure 14. Summary of results for $I_{d,max}$, impulse bit and discharge energy W_d for varying battery voltage U_b.

One of the most essential problems for future high-frequency MPC applications is the erosion of the electrode system. Figure 15 shows the results of the erosion process after approximately 10^3 main discharge ignitions with a maximum current no more than 30 kA. The observed level of erosion was unexpectedly low as the studied high-current arc discharges with the self-induced strong tangential magnetic fields have fast moving cathode spots. But the main visual problem is in the diverter area (hole on the center of cathode) as shown in Figure 15. Possible ways for erosion decreasing in this area can be reached by changing the diverter diameter and shape. The theory and simulation of the plasma behavior in the MPC diverter area is very complicated. So future magneto-plasma simulations and experimental trials might give indications for a better diverter geometry with less erosion.

Nevertheless, future air-breathing magneto-plasma propulsion systems will need up to 10^3 ignitions per second. So alternative materials from fusion reactors, amorphous metals and special alloys with different porosity and surface structuring will be be also investigated in the future. In this regards, the MPC itself might be useful to manipulate the surface structure of future electrode materials [7].

Figure 15. Results of erosion processes in MPC after 10^3 launches.

In future experiments the plasma dynamics could be investigated using ultrafast cameras with up to 2 Mio frames per seconds which are available to the corresponding authors. In the present work the maximum voltage and power limits of the new MPC thruster were not tested. The main task was the first demonstration of a pulsed MPC-based plasma thruster with ns-internal excitation for a stabile operation at high atmospheric pressures up to 1 bar. In this regard, a first breakthrough and pulse operation with 4.7 Hz was demonstrated.

In the next step, the pulse frequency of the main discharge will be increased up to 10 Hz. Furthermore, a new mobile power generator will be developed for the first flight demonstration onboard of the b-Ionic Airfish, which was the world's first airship propelled by plasma engines in 2005, see Figure 16a-16b. Only 0.08 N or 8 g would be sufficient to propel this 7.5 m airship at low speeds up to 1 m/s [32]. A 5 Hz thruster has already about 0.02 N. So an array of four plasma pulse "detonation" thrusters with the present power level would make it fly. The available maximum weight for the power generator is about 5.1 kg plus 1.2 kg for LiPo batteries [32].

The general thrust of an array with 10 cells, each operating with a pulse frequency of 50 Hz, is 2.0 N. The total power required for a first high altitude (H=20 km) demonstrator mission using an array with 10 thruster is about 75 kW. With a solar battery effectiveness of 0.2-0.3, the required minimum surface area is 250 m².

In any case, the new propulsion technology is still away from being competitive but it has to be noted that the research, development and optimization is now at the very beginning. The impulse bit for each thruster unit can be essentially increased by using different ejector schemes and jet focusing nozzle structures. These are items of next investigations.

Furthermore, there are also a large amount of other possible technological applications in the field of aerodynamics, material sciences and power engineering. But a real flight demonstration is the next milestone goal towards new magneto-plasma flux compression thrusters for stratospheric airships or high altitude platform stations (HAPS) which are currently all limited to about 25 km altitude by using propellers. With future air-breathing magneto-plasma flux compression thrusters next generation solar, beamed or fusion energy powered airships could climb to altitudes up to 50 km and beyond.

(a)

(b)

Figure 16a and 16b. The world's first plasma propelled airship, the 7.5 m long b-Ionic Airfish (2005) [32].

Acknowledgements
The breakthrough experiments were funded by Electrofluidsystems and performed by the authors in the company's future workshop in Berlin. The numerical studies were performed by Tatiana Banokina in St. Petersburg funded by a grant of the St. Petersburg State University with number 11.37.167.2014.

References

1. Morozov A 1975 Processes in a magnetoplasma compressor. *Physics of Plasma* **1** 95-101 (in Russian).
2. Astashinsky B, Bakanovich G, Minko L 1980 Investigations of dynamic of plasma arising and forming processes of compression area in gas magneto-plasma compressor. *Journal of Applied Spectroscopy* **33** 629-633 (in Russian).
3. Ananin S, Astashinsky B, Bakanovich G, Kostukevich E, Kusmitsky A, Manykovsky A, Minko L, Morozov A 1990 Investigations of processes of plasma jets rising in quasi stationary high-current plasma accelerator (QHPA). *Physics of Plasma* **16** 186-196 (in Russian).
4. Morozov A 1990 The principles of coaxial quasi stationary plasma accelerators. *Physics of Plasma* **16** 131-146 (in Russian).
5. Astashinsky B, Bakanovich G, Kusmitsky A, Minko L 1992 Choosing of operation modes and plasma parameters of magneto plasma compressor. *Engineering Physical Journal* **62** (3) 386-390 (in Russian).
6. Astashinsky B, Bakanovich G, Minko L 1980 Investigations of dynamic of plasma arising and forming processes of compression area in gas magneto-plasma compressor. *Journal of Applied Spectroscopy* **33** 629-633 (in Russian).
7. Kuraica M M, Astashynski V M, Dojcinovic I P, Puric J 2002 Modification of solid surface by a compression plasma flow. In *Physics of Laser Crystals*, edited by Krupa J-C, Kulagin N A, NATO Science Series II. Mathematics, Physics and Chemistry **126** 245-254.
8. Mashek I, Anisimov Yu I, Lashkov V A, Kolesnichenko Yu F 2006 Quasi-stationary magneto-plasma compressor for investigation of plasma jets in aerodynamics. 44th Aerospace Science Meeting, Reno, AIAA-2006-1458.
9. Mashek I Ch, Anisimov Yu I, Lashkov V A, Kolesnichenko Yu F 2007 Investigation of plasma jets generated by quasi-stationary magneto-plasma compressor under the high static pressure. 45th AIAA Aerospace Sciences Meeting and Exhibit, Reno, AIAA 2007-0221.
10. Mashek I Ch, Anisimov Yu I, Lashkov V A, Kolesnichenko Yu F 2008 Plasma actuator for supersonic flow based on miniature magneto-plasma compressor. 46th AIAA Aerospace Sciences Meeting and Exhibit, Reno, AIAA 2008-1408.
11. Mashek I Ch, Lashkov V A, Kolesnichenko Yu F, Brovkin V G 2011 Investigation of miniature magneto-plasma compressor as a basic part for perspective plasma actuators for supersonic flows. 49th AIAA Aerospace Sciences Meeting Including the New Horizons Forum and Aerospace Exposition, AIAA-2011-1274.
12. Mashek I, Lashkov V, Vinerov I, Mikhailov D 2012 Temperature dynamics of plasma focus in air-breathing magneto-plasma thruster. 11th International Workshop on Magneto-Plasma Aerodynamics, JIHT RAS Moscow.
13. Mashek I Ch, Lashkov V A, Göksel B, Paschereit O, Tajmar M 2012 Investigation of magneto-plasma compressors with internal initiation to develop high momentum pulsed plasma jet actuators for flow control. *Conference Proceedings of the Joint ERCOFTAC/PLASMAERO Workshop,* Paper 32, 108-109.
14. Göksel B, Mashek I, Lashkov V et al 2013 Novel air-breathing plasma-jet propulsion for solar powered high altitude flight platforms. EUCASS, Munich, Paper a593.
15. Trklja N 2015 Current status of the magnetoplasma compressor device in Belgrade – study of plasma facing materials important for fusion reactors. *Interdisciplinary Description of Complex Systems* **13** (1) 173-181.
16. Repin P B, Rep'ev A G 2001 Self-organization of the channel structure of a nanosecond diffuse discharge in a wire-plane electrode system. *Technical Physics* **46** (5) 632-634.
17. Rakitin A E, Zhukov V P, Starikovskii A Yu 2009 Deflagration-to-detonation transition under initiation by high-voltage nanosecond discharges. *Progress in Propulsion Physics* **1** 367-378.
18. Bak M S, Kim W, Cappelli M A (2011) On the quenching of excited electronic states of molecular nitrogen in nanosecond pulsed discharges in atmospheric pressure air. *Applied Physics Letters* **98**, 011502-1 – 011502-3.
19. Cheng D Y 1970 Plasma deflagration and the properties of a coaxial plasma deflagration gun. *Nucl. Fusion* **10** (3) 305.
20. Poehlmann F, Gascon N, Cappelli M 2007 The deflagration-detonation transition in gas-fed pulsed plasma accelerators. 43rd AIAA/ASME/SAE/ASEE Joint Propulsion Conference & Exhibit, Cincinnati, AIAA 2007-5263.

21. Loebner K T K, Wang B C, Poehlmann F R, Watanabe Y, Cappelli M A 2014 High-Velocity Neutral Plasma Jet Formed by Dense Plasma Deflagration. *IEEE Transactions on Plasma Science* **42** (10) 2500-2501.
22. Loebner K T K, Underwood T C, Cappelli M A 2015 Evidence of Branching Phenomena in Current-Driven Ionization Waves. *Physics Review Letters* **115** (17) 175001-175005.
23. Pollard W, Carpenter B, Duggleby, A, Staack, D 2010 Design and preliminary characterization of a micro-scale dense plasma focus. 52nd Annual Meeting of the APS Division of Plasma Physics, Abstract DPPXP9056P.
24. Pollard W, Duggleby A, Staack A 2016 Dynamics of two microscale DPF devices. *Journal of Physics D Applied Physics* **49** (5) 055201.
25. Wang Z P, Yousefi H R, Nishino Y, Ito H, Masugata K 2009 Fabrication of DLC films by pulsed ion beam ablation in a dense plasma focus device. *Physics Letters A* **373** 4169-4173.
26. Haruki T, Yousefi H R, Sakai J-I 2010 Simulation of plasma heating caused by the coalescence of multiple loops in a proton-boron fusion plasma. *Physics of Plasmas* **17** (3) 032504.
27. Gribkov V A 2008 Current and perspective applications of dense plasma focus devices. *AIP Conf. Proc.* **996** 51.
28. Schmidt A, Link A, Welch D, Meehan T, Tang V, Halvorson C, May M J, Hagen E C 2014 Fully kinetic simulations of megajoule-scale dense plasma focus. *Physics of Plasmas* **21** (10) 102703.
29. Polsgrove T, Fincher S, Adams R B, Cassibry J, Cortez R, Turner M, Maples C D, Miernik J N, Statham G N, Fabisinski L, Santarius J, Percy T 2011 Design of z-pinch and dense plasma focus powered vehicles. AIAA Aerospace Sciences Meeting, Orlando, AIAA 2011-962.
30. Thomas R, Yang Y, Miley G H, Mead F B 2005 Advancements in dense plasma focus (DPF) for Space Propulsion. *AIP Conf. Proc.* **746**, 536.
31. Knecht S D, Thomas R E, Mead F B, Miley G H, Froning D 2006 Propulsion and power generation capabilities of a dense plasma focus (DPF) fusion system for future military aerospace vehicles. *AIP Conf. Proc.* **813**, 1232.
32. Göksel B, Fischer M, Rechenberg I, Thallemer A 2005 Electrostatic plasma wave propulsion for bionic airships (Elektrostatischer Plasma-Wellantrieb für bionische Luftschiffe). *Proceedings of the German Aerospace Congress 2005*, **3** 1853-1856 (Paper DGLR-2005-261 in German).

Titanium nitride plasma-chemical synthesis with titanium tetrachloride raw material in the DC plasma-arc reactor

D.E.Kirpichev, M.A.Sinaiskiy, A.V.Samokhin, N.V.Alexeev

A.A.Baikov Institute of metallurgy and material science, Moscow, Leninskiy av., 49

dym2004@bk.ru, samokhin@imet.ac.ru

Abstract. The possibility of plasmochemical synthesis of titanium nitride is demonstrated in the paper. Results of the thermodynamic analysis of $TiCl_4 - H_2 - N_2$ system are presented; key parameters of TiN synthesis process are calculated. The influence of parameters of plasma-chemical titanium nitride synthesis process in the reactor with an arc plasmatron on characteristics on the produced powders is experimentally investigated. Structure, chemical composition and morphology dependencies on plasma jet enthalpy, stoichiometric excess of hydrogen and nitrogen in a plasma jet are determined.

1. Introduction

Titanium nitride possesses unique physicochemical and physicomechanical properties [1, 2]. This results in a wide range of applications on TiN basis, such as production of the cutting tools with wear proof coatings. The newly developed nanomaterials processing techniques can improve the resulting wear properties in comparison with traditional techniques. Thus there is a great interest in nano-powders based on inorganic compounds, including titanium nitride. Processes using low-temperature plasma for production of titanium oxygen-free compound powders are currently used [3-5]. Plasma synthesis is the most universal method for nano-powder production for various size and chemical compositions. Nevertheless, information on commercial realization of plasma processes of titanium nitride nano-powders production is absent.

For practical realization of titanium nitride nano-powder synthesis the most common raw material is the titanium tetrachloride, which is commercially used for titanium and titanium dioxide production. Titanium tetrachloride has a low boiling temperature (410 K) and can be easily transferred to the gaseous state, providing possibility of plasma synthesis with participation of other gaseous reagents and production of uniformly graded nano-powders.

Practical realization of nanopowder production in thermal plasma requires an equipment design, which provides the necessary processes efficiency and has along service life. Plasma reactors on the basis of arc plasmatrons are known to be the most efficient in this field. Today an plasma-arc gas heater (plasmatron) is one of the most widespread devices for low-temperature plasma generation [8, 9]. Its advantages are the possibility of heating of any gases and mixes to rather high average mass temperatures (1000-5000 K), effective efficiency of heating (up to 90%), a high resource of long-term operation life (up to 1000 h), relatively simple of experimental installations design, sufficient ease in management of operating modes at simultaneous high reliability and stability.

The main goal of this work was the realization of titanium nitride nano-powders synthesis in the laboratory scale with an option for future commercialization. For this purpose the synthesis of target products based the interaction of titanium tetrachloride vapor (or its mixes with methane) with the hydrogen-nitrogen thermal plasma generated in the arc plasmatron in the reactor with a limited jet current was chosen.

Possibility of process commercialization depends on the existence of a raw material source, the existing technical solutions for effective generators of thermal plasma working in the wide power range and also design of the plasmochemical reactor with a long service life.

2. Thermodynamic analysis

It is necessary to define the reaction products and titanium nitride output quantity in the reaction of titanium tetrachloride with nitrogen in the hydrogen at high temperatures and the process power consumption. In the presented work calculations of equilibrium structures and thermodynamic properties of $TiCl_4 - H_2 - N_2$ system are performed. For these calculations TERRA program complex for modeling of phase and chemical equilibrium in multicomponent systems is used [6, 7]. For the calculations it is assumed that the reactions take place in isobaric and isothermal conditions at temperatures in the range from 400 to 4000 K and a pressure of 0.1 MPa. Further calculation parameters are given in Table 1.

Table 1. Main parameters and relations of calculations of equilibrium structures and thermodynamic properties of $TiCl_4 + x\ H_2 + y\ N_2$ system

x	y	H/Cl	N/Ti
8	0.5	4	1
8	5	4	10
8	20	4	40
20	0.5	10	1
20	5	10	10
20	20	10	40

Based on the calculations results of equilibrium compositions and thermodynamic properties of $TiCl_4 + xH_2 + yN_2$ reactions products the following equilibrium characteristics of processes were defined:

- titanium nitride output TiN,

$$\{TiN_{output}\} = \frac{[TiN]}{Ti_{sum}}, \quad (1)$$

with [TiN] as titanium nitride content, mole/kg,
Ti_{sum} – total titanium content in the system, mole/kg,

- titanium chlorides $TiCl_m$ output,

$$\{TiCl_{m\,output}\} = \frac{[TiCl_m]}{Ti_{sum}}, \quad (2)$$

with $[TiCl_m]$ as titanium chloride content, mole/kg,

- energy consumption necessary for 1 kg of titanium nitride produce at the set temperature of mixture, MJ/kg,

$$\{Energy\ consumption\} = \frac{(I_{Tsum} - I_{0sum})}{[TiN]*Mr(TiN)}, \quad (3)$$

where I_{Tsum} - total full enthalpy of equilibrium mixture at temperature T, kJ/kg,

IOP Conf. Series: Journal of Physics: Conf. Series **825** (2017) 012006 doi:10.1088/1742-6596/825/1/012006

I_{0sum} – total full enthalpy of initial reagents ($TiCl_4$ temperature is 500 K, other reagents – 298 K), kJ/kg, and Mr(TiN)– TiN molecular weight;

- enthalpy of nitrogen-hydrogen plasma, which provides equilibrium mixture at the set temperature,

$$\{\text{Plasma enthalpy}\} = \frac{(I_{Tsum} - I_{0sum})}{(x+y)\cdot 22.4}\cdot 1000, \quad (4)$$

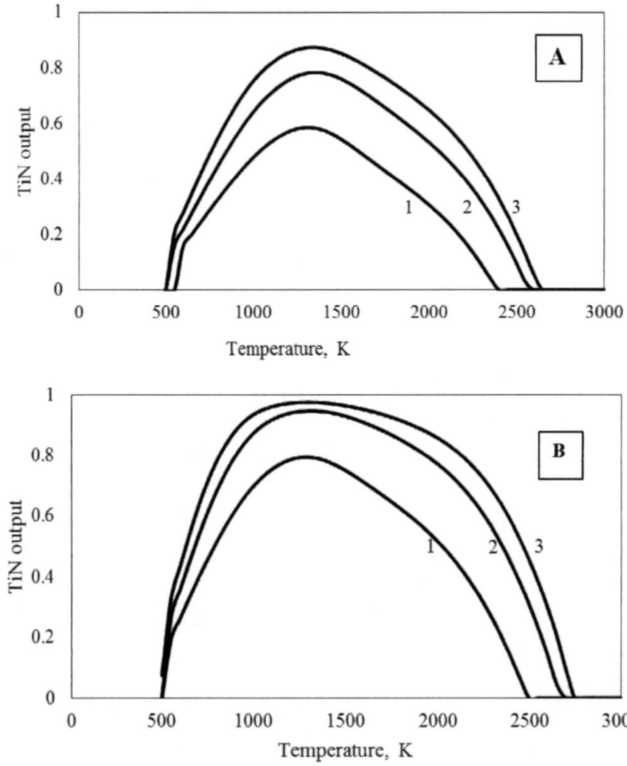

Figure 1. Dependency of TiN output on temperature at various compositions of initial reagents for (A) H_2 ratio x=8 and N_2 ratio (1) y=0.5, (2) y=5; (3) y=20 and (B) H_2 ratio x=20 and N_2 ratio (1) y=0.5, (2) y=5; (3) y=20.

As a result of the calculations dependencies of TiN output on temperature and a ratio of initial reagents are established (figure 1). It is calculated that values above 95% of TiN output are provided at $x = 20$ and at $y \geq 5$ and temperature 1300 – 1400 K.

At a temperature exceeding 1400 K there is a decrease in titanium nitride output. Thus in the whole system there is a formation of mainly $TiCl_3$ and $TiCl_2$ up to the temperature of 3000 K (figure 2). At a temperature below 1300 K titanium nitride transforms to tetrachloride $TiCl_4$.

It follows from calculations that in the equilibrium conditions during the synthesis of titanium nitride the main impurity will be $TiCl_3$ and $TiCl_4$. These components have a boiling temperature of 1230 and 410 K [10]. In order to decrease the concentration of these impurities in the product thermal vacuum processing can be used, considering that titanium nitride is a refractory compound.

IOP Conf. Series: Journal of Physics: Conf. Series **825** (2017) 012006 doi:10.1088/1742-6596/825/1/012006

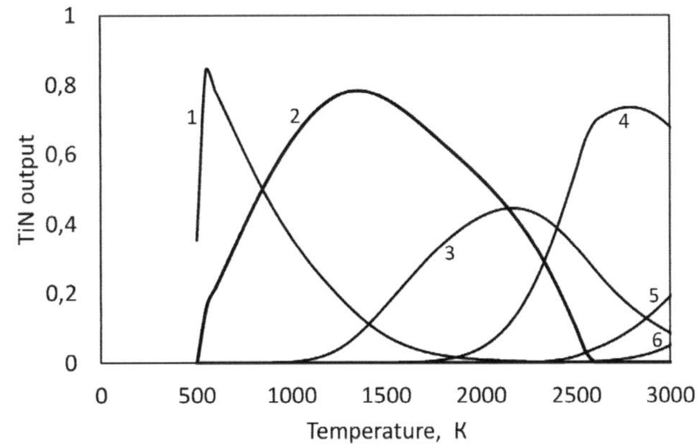

Figure 2. Dependence of titan ferrous components output on temperature at x=8, y=5. 1 – $TiCl_4$, 2 – TiN, 3 – $TiCl_3$, 4 – $TiCl_2$, 5 – TiCl, 6 – Ti

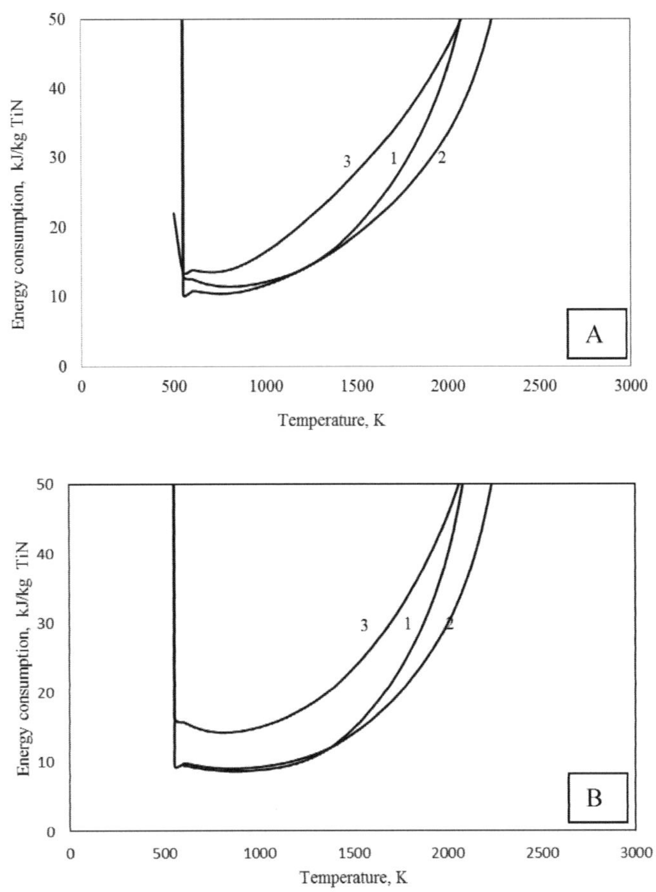

Figure 3. Dependence of energy consumption of titanium nitride produce on temperature at various compositions of initial reagents (A) H_2 ratio x=8 and N_2 ratio (1) y=0.5, (2) y=5; (3) y=20 and (B) H_2 ratio x=20 and N_2 ratio (1) y=0.5, (2) y=5; (3) y=20.

In the temperature field of the maximum titanium nitride output the energy consumption of its synthesis is 16 – 20 MJ/kg depending on excess of nitrogen (figure 3). The absolute minimum of energy consumption is in the temperature range of 550 – 750 K (at x=20) and 800 – 900 K (at x=8). However in these conditions titanium nitride output does not exceed 55 – 70%.

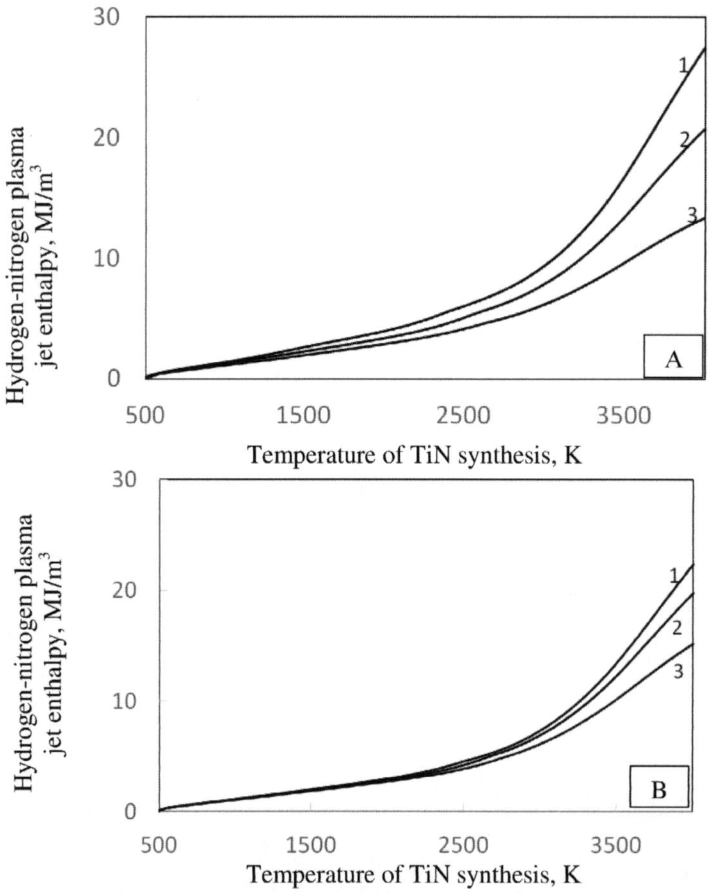

Figure 4. Dependence of hydrogen-nitrogen plasma jet enthalpy on TiN process synthesis temperature at various compositions of initial reagents (A) H_2 ratio x=8 and N_2 ratio (1) y=0.5, (2) y=5; (3) y=20 and (B) H_2 ratio x=20 and N_2 ratio (1) y=0.5, (2) y=5; (3) y=20.

For providing the maximum titanium nitride output at x=20 and at y≥0.5 (process temperature - 1300 K) the necessary value of nitrogen-hydrogen plasma enthalpy is about 1.5 MJ/m^3 (figure 4). This enthalpy can be easily provided and exceeded when using the existing designs of arc plasmatrons.

3. Experimental setup

Experimental investigations of titanium nitride nano-powder synthesis were conducted using a plasma setup based on the thermal plasma arc generator (figure 5). In this setup a power of 25 kW is necessary to provide interaction in the titanium tetrachloride - nitrogen mixture with hydrogen-nitrogen-argon thermal plasma.

In the reactor chamber a mixture of initial reagents and their chemical transformations used for formation of TiN nanoparticles are present. The nanoparticles were accumulated on reactor walls, and also were taken out on the bag filter. After the filter gaseous products of reaction come to a bubbling absorber where there absorption of chlorine-containing products takes place.

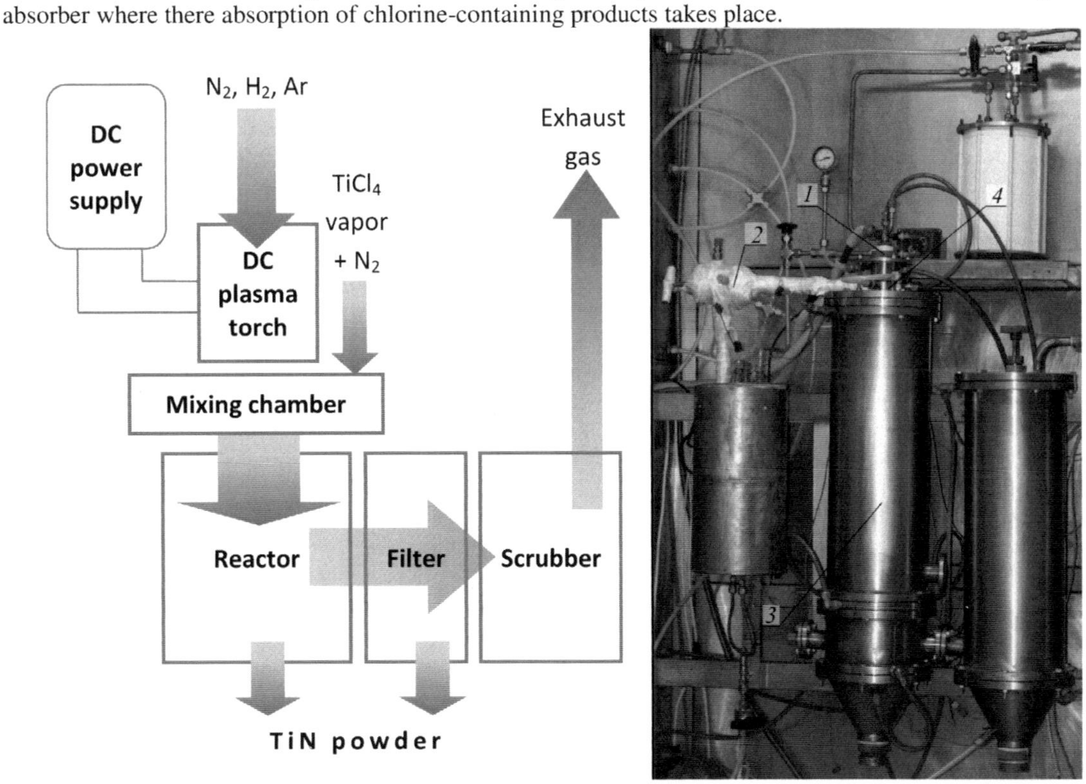

Figure 5. Experimental setup.
1 – DC plasma torch, 2 – $TiCl_4$ + N_2 steam line, 3 – reactor, 4 – mixing chumber

All heat loaded setup units are cooled with water. The cooling water thermal losses in each unit were defined by calorimetry. The heat of Q_i determined by a calorimetry which is taken away with water from a unit i (thermal losses) was calculated as:

$$Q_i = \frac{(T_i - T_0)\cdot G_i}{860}, \text{ [kW]} \tag{5}$$

where T_i – water temperature at the exit from unit, °C; T_0 – water temperature on an entrance, °C; G_i – consumption cooling a water unit, kg/h.

The enthalpy of a plasma jet I_{pl} was calculated as the difference between the power N brought from a DC power source to a plasmatron and thermal losses Q_{pl} in plasmatron related to a total consumption of plasma-forming gases ΣVj:

$$I_{pl} = \frac{N - Q_{pl}}{\sum_j V_j}, \text{ [kWh/n.m}^3\text{]}. \tag{6}$$

Ratio of hydrogen of plasma and chlorine in $TiCl_4$ and ratio of nitrogen of plasma and titanium in $TiCl_4$ were set by change of a ratio of titanium tetrachloride and plasma-forming gases consumptions.

Liquid titanium tetrachloride fed the dosing pump in a high-temperature zone of the evaporator. Titanium tetrachloride vapor was injected by the transporting gas at a temperature of 200 °C in a plasma jet. As the transport gas nitrogen was used.

IOP Conf. Series: Journal of Physics: Conf. Series **825** (2017) 012006 doi:10.1088/1742-6596/825/1/012006

The complex physical and chemical analysis of the produced nanopowders included:

- the X-ray phase analysis (RFA) which is carried out on the RIGAKU Ultima-4 diffractometer in the filtered Cu Kα radiation with the high-speed detector "D/teX", a software package of "PDXL" and the databank "PDF-2";
- measurement of powders specific surface by BET method on the analyzer of a specific surface and porosity "Micromeritics TriStar 3000";
- measurement of particle size distribution of the produced powders by laser diffraction method "Mastersizer 2000 M";
- electron microscopy on the scanning microscope (SEM) "JSM-6700F" ("Jeol Company") and on the translucent microscope (TEM) "Tecnai G2 F20" ("FEI Company");
- measurement of the total nitrogen content on the LECO TS-600 analyzer;
- measurement of the total chlorine content by Sheniger's method.

Produced TiN powder mean particles size depends on process parameters. Distribution by the size is normal for each operation. TiN powders dispersion changed in the range 230 – 30 nm, it includes the range of nanopowders with the size below 100 nm.

Main parameters of plasma process and the analyzed physical and chemical properties of the produced nano-powders TiN_x are specified in table 2.

Table 2. Main parameters of plasma process and the analyzed physical and chemical properties of the produced nanodimensional powders TiN_x

parameter	range	dimension
Mass rate of flow $TiCl_4$, G_{TiCl4}	0,1 – 0,5	kg/h
Concentration Ti, C_{Ti}	3,4 – 27,1)·10 3	mol Ti/mol gas
H_2 consumption in plasma, G_{H2}^{pl}	0,2 – 1,4	$n.m^3/h$
N_2 consumption in plasma, G_{N2}^{pl}	0,2 – 0,8	$n.m^3/h$
Ar consumption in plasma, G_{Ar}^{pl}	0,0 – 1,6	$n.m^3/h$
N_2 consumption in transport gas, G_{N2}^{tr}	0,0 – 1,1	$n.m^3/h$
Temperature of powder accumulation surface, T_{qz}	401 – 815	°C
Average integrated temperature of powder accumulation surface, T_{qz}^{av}	557 – 687	°C
Plasma jet enthalpy, I_{pl}	3,0 – 6,3	kWh/m^3
Ratio of hydrogen of plasma and chlorine in $TiCl_4$, $H_{pl}/Cl = (2\ G_{H2}^{pl}/22,4)/(4\ G_{TiCl4}/Mr(TiCl_4))$	2 – 47	mole/mole
Ratio of nitrogen of plasma and titanium in $TiCl_4$, $N_{pl}/Ti = (2\ G_{N2}^{pl}/22,4)/(4\ G_{TiCl4}/Mr(TiCl_4))$	10 – 129	mole/mole
Specific surface, Ssp	5 – 39	m^2/g
Medium particle size, $d = 1103,24/S_{sp}$	28 - 234	nm
Nitrogen mass concentration, [N]	5,1 – 22,5	mass.%
Chlorine mass concentration, [Cl]	<0,10 – 4,45	mass.%

4. Results and discussion

It is experimentally established that at interaction of $TiCl_4$ vapors with hydrogen-nitrogen plasma leads to formation of titanium nitride powder. According to the X-ray phase analysis (figure 6) the produced single-phase nano-powders also have a cubic crystal lattice like NaCl with the total ratio of chlorine of ~0.1%. This was provided due to the consumption of titanium tetrachloride of 0.2 kg/h. The resulting particle size lies in a wide range and depends on process parameters.

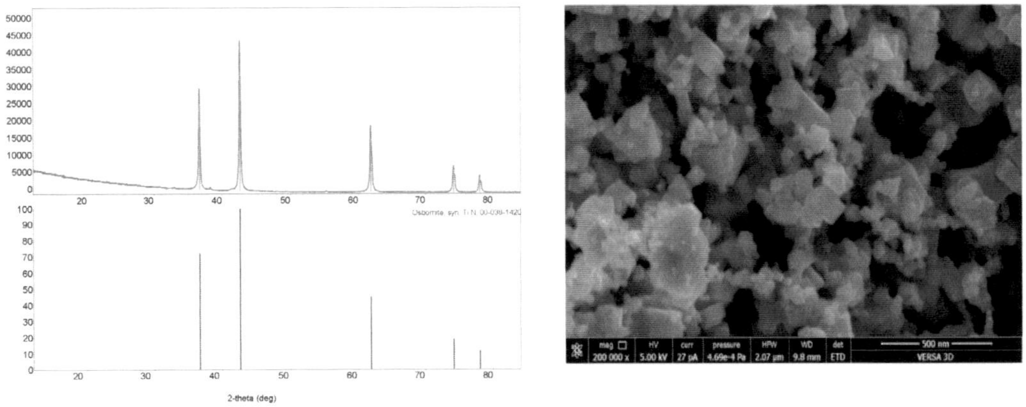

Figure 6. X-ray and SEM results for TiN nano-powders

The increase in consumption of raw materials to 0.4 kg/h led to decrease in a target product output and increase in the total content of chlorine to ~1%. Thus on a x-ray diffraction pattern the phase of the hydrolyzed titanium trichloride $TiCl_3 \cdot 6H_2O$ was identified. Presence of this phase is caused by sorption of $TiCl_3$ on surface of titanium nitride nanoparticles and their interaction with the water vapors, which are present in the surrounding atmosphere during operation of removal of nano-powder from the wall of the reactor performed in air.

In the produced titanium nitride nanopowders the content of nitrogen changed in the range of 18.8-22.5 mass.%. That corresponds to formulas $TiN_{0.79}$-$TiN_{0.99}$ and is within the range of homogeneity of titanium nitride. Results of SEM investigations has shown that all produced powders represent ensembles of nanoparticles mainly of cubic form with sizes of 20-150 nanometers and aggregates on their basis (figure 7, A). In nano-powders in which presence of $TiCl_3 \cdot 6H_2O$ was detected, change of their morphology is noted – mainly facetted form of particles was replaced with roundish that is caused by formation of a cover of $TiCl_3 \cdot 6H_2O$ on originally cubic nanoparticles of titanium nitride (figure 7, B).

Figure 7. TiN (A) and with $TiCl_3 \cdot 6H_2O$ (B) presence nanopowders produced SEM results

Formation of $TiCl_3 \cdot 6H_2O$ cover of titanium nitride nano-powder is undesirable. Therefore removal of the produced titanium nitride from the reactor has to be carried out without contact with air.

The specific surface of titanium nitride nano-powders varied within 11–39 m^2/g. That corresponds to the range of the average size of particles of 100-28 nm. This value is calculated from ratio $d=6/(\rho_{TiN} \cdot S)$, where d is the average size of particles, ρ_{TiN} density of nitride of the titan and S the size of a specific surface.

The energy level at which the process was performed is defined by a plasma jet enthalpy, which defines distribution of temperatures in the volume of the reactor. It is one of the major factors influencing both physical and chemical properties of the produced nano-powders and characteristics of the process, such as conversion degree of initial reagents, a target product output and the electric power consumption.

As a result of the executed experiments it is established that change of an enthalpy of a plasma jet has the strongest impact on particle size of the produced titanium nitride. With growing plasma jet enthalpy the specific surface of the produced titanium nitride nano-powder decrease is observed. Thus, the average size of nanoparticles calculated from the measured value of a specific surface increases (fig. 8).

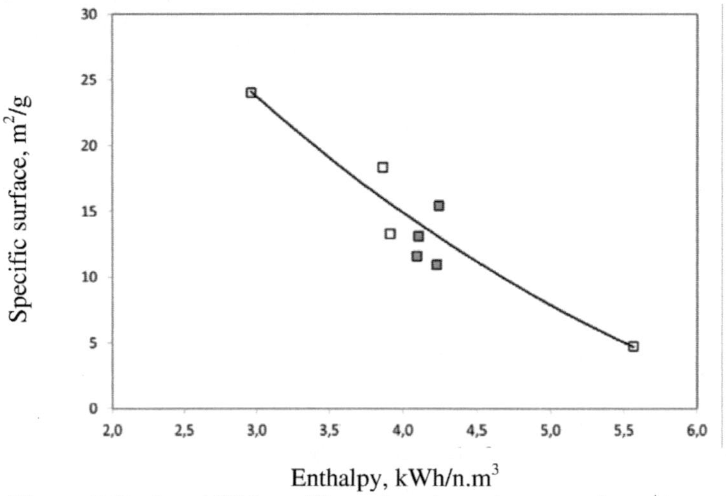

Figure 8. Produced TiN specific surface dependence on plasma jet enthalpy

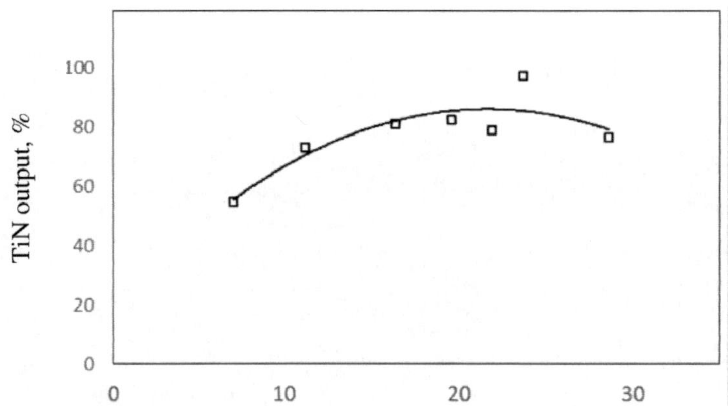

Figure 9. Produced TiN output dependence on molar ratio H/Cl

For a consumption of titanium tetrachloride of 0.2 kg/h the increase in enthalpy of a plasma jet of 3.0-5.6 kWh/m^3 leads to decrease of a specific surface 24.0–4.7 m^2/g. This corresponds to change of the average size of particles of 46-230 nm. Existence of a facet of particles indicates that titanium nitride nanoparticles in plasma process is formed on the “vapor - crystal” macromechanism, The increase in the average size of particles with growth of the enthalpy of the plasma jet suggests that the prevailing growth rate of particles is related to the speed of their formation.

The phase structure of the received products remained invariable when increasing the plasma jet enthalpy and the average size of particles, respectively. However it led to some decrease in the content of nitrogen in the produced nano-powder. At maximum enthalpy of 5.6 kWh/m^3 the content of nitrogen is 18.8 mass.%. Change of a molar ratio of the H/Cl elements within a range of 10-30 had no essential impact on the size of a specific surface of nano-powder of titanium nitride, however, influenced its output (figure 9).

The increase in the H/Cl ratio in the range of 7-20 led to growth of titanium nitride output. When further increasing the ratio, the output decreased a little. Existence of an extremum in the considered dependence is caused by influence of opposite operating factors – growth of a ratio of H/Cl (surplus of a reducer) leads to increase of the output. But at the same it time reduces process temperature at the expense of increase in amount of the hydrogen entering the reactor. Therefore the considered dependence has a maximum.

5. Conclusions

The presented investigations have shown the possibility of synthesis of titanium nitride nano-powders from vapors of titanium tetrachloride in a stream of the hydrogen-nitrogen plasma generated in an arc plasmatron.

From results of the carried-out thermodynamic calculation of equilibrium mixture can be concluded, that providing the target products output close to 95% requires tenfold excess of hydrogen in comparison to the stoichiometrically necessary amount, thus by-products of synthesis are the lowest titanium chlorides. Dependence of all target products output on temperature has extreme character and a maximum of titanium nitride output are obtained for the process temperature in range of 1100-1500K. The minimum energy consumption for production of 1 kg of target products in the vicinity of their maximum output is 20 MJ/kg.

The target nano-powders output reached in experiment is up to 94%, thus the total content of impurity of chlorine could be minimized to less than 0.1 mass.%.

In experiments single-phase nanopowders with a cubic crystal lattice like NaCl and the representing ensembles of nanoparticles of mainly cubic form with sizes of 20-150 nm and aggregates on its basis are produced. Change of parameters of synthesis allowed producing the titanium nitride nano-powders with a specific surface in the range of 11-39 m^2/g containing 18.8–22.5 mass.%. This corresponds to gross - formula $TiN_{0.79}$-$TiN_{0.99}$.

Further investigations have to be performed in order to optimize the process, which can provide the maximum output of target products and the minimum consumption of the electric power when receiving target products with the set of physical and chemical properties.

This research was conducted with support of the Russian Ministry of Education and Science (Federal Target Program «Research and development on priority directions of scientific-technological complex of Russia for 2014 - 2020 years», project «Development of bases of plasma-chemical technologies for production of nanosized powders of titanium anoxic compounds - nitride, carbide and carbo-nitride for developing of new structural and functional materials», agreement № 14.607.21.0103, unique code RFMEFI60714X0103).

References

[1] Pierson Hugh O. Handbook of refractory carbides and nitrides: Properties, characteristics, processing and applications. William Andrew Publ., 1997, 362 p.

[2] Shackelford J.F., Alexander W. CRC materials science and engineering handbook. CRC Press, 2000, 1980 p.

[3] Mahendra Kumar S., Murugan K., Chandrasekhar S.B., Neha Hebalkar, Krishna M., Satyanara B.S., Giridhar Madras. Synthesis and characterization of nano silicon and titanium nitride

powders using atmospheric microwave plasma technique. J.Chem.Sci., 2012, v.124, No.3, p.557-563.

[4] Alekseev N.V, Samokhin A.V., Tsvetkov Yu.V. Synthesis of titanium carbonitride nanopowder by titanium tetrachloride treatment in hydrocarbon-air plasma. High Energy Chemistry, 1999, v.33, No.3, p.194-197.

[5] Poluchenie ultradispersnykh nitridov v plazme SVCh-razryada [Production of ultradisperse nitrides in microwave plasma]. In: Batenin V.M., Klimovsky I.I., Lysov G.V., Troitsky V.N. SVCh-generatory plazmy: fizika, tekhnika, primenenie [Plasma microwave generators: physics, technical equipment, application]. Moscow: Energoatomizdat Publ., 1988, p.178-213. (In Russ.).

[6] Vatolin N.A., Moiseev G.K., Trusov B.G. Termodinamicheskoe modelirovanie v vy`sokotemperaturny`kh neorganicheskikh sistemakh [Thermodynamic modeling of high temperature nonorganic systems]. Moscow, Metallurgia, 1994, 352 p. (In Russ.)

[7] Trusov B.G. Komp`iuternoe modelirovanie fazovy`kh i himicheskikh ravnovesii` [The program system of modeling of phase and chemical equilibriums]. ISSN 0236-3933, Vestnik MGTU im. N.E. Baumana, ser. "Priborostroenie" , 2012, p. 240 -249 (In Russ.)

[8] Tumanov Yu.N. Plazmennye, vysokochastotnye, mikrovolnovye i lazernye tekhnologii v khimiko-metallurgicheskikh protsessakh [Plasma, high-frequency, microwave and laser technologies in chemical and metallurgical processes]. Moscow: Fizmatlit Publ., 2010, 968 p. (In Russ.)..

[9] Zhukov M.F., Zasy`pkin I.M., Timoshevskii` A.N. E`lektrodugovy`e generatory` termicheskoi` plazmy` [Electric arc generators of thermal plasma] Novosibirsk: Nauka, 1999, 712 p

[10] Perry D.L. Handbook of inorganic compounds. CRC Press, 1995, 578 p.

IOP Conf. Series: Journal of Physics: Conf. Series **825** (2017) 012007 doi:10.1088/1742-6596/825/1/012007

Plasma actuators for active flow control based on a glow discharge

M. Kühn[1], M. Kühn-Kauffeldt[1], J. Schein[1], A. Belinger[2]
[1]Laboratory For Plasma Technology, University Of The Federal Armed Forces Munich, Germany
[2] LAPLACE, Universit de Toulouse, CNRS, INPT, UPS, France
E-mail: marvin.kuehn@unibw.de

Abstract. In this work a glow discharge based active flow control for high flow velocities and low Reynolds numbers is presented. Unlike common plasma actuators such as dielectric barrier discharge (DBD) or spark jets, this actuator uses small impulse bits at frequencies. The actuator is optimized for frequencies up to 40 kHz to counter Tollmien Schlichting wave effects and so reduce overall air foil drag. Several measurements to prove the non-eroding effect of the actuator and the electrical properties were performed. It was found that the actuator is capable of operating at high frequencies without measurable erosion.

1. Introduction

Nowadays strict laws against air pollution require more efficient aircrafts in order to fulfill environment protection regulations. This can be achieved by introducing actuators on the wing surface, which are able to decrease the flow induced drag and thereby to lower the overall energy consumption. Various passive and active mechanical systems have been developed for this purpose. For example fluid oscillation actuators are in use for turbine blades for several years and could demonstrate that they can improve the flow characteristic of modern turbines [1].

Besides mechanical actuators electrically driven plasma actuators can also induce flow perturbations. Their main advantage is that the strength and the frequency of perturbation can be electrically controlled, introducing a flexible system, applicable for a whole variety of problems. Actuators either can be used as a tool for investigation turbulent instabilities or as a system which is able to react to a big range of instabilities. In the last decade, different types of plasma actuators for various flow control applications have been developed and investigated [2, 3, 4]. Low Pressure Turbines (LPT) are one of the applications for which flow control can improve the overall performance by influencing the flow induced drag and flow separation. In an LPT the energy loss is induced by Tollmien-Schlichting Waves (TSW). As illustrated in fig. 4(top) TSW break up the laminar flow and create high frequency turbulences at high aerodynamic loads. Active flow actuators in the TSW evolution zone can keep the flow more stable and reduce the drag (fig. 4 bottom).

First studies at the University of the Federal Armed Forces of Germany in Munich with a cathodic arc discharge showed a too high temperature stress, eroding the cathode surface (arc root points), the carrier plate (heat transfer) and creating a passive actuator. Also the power

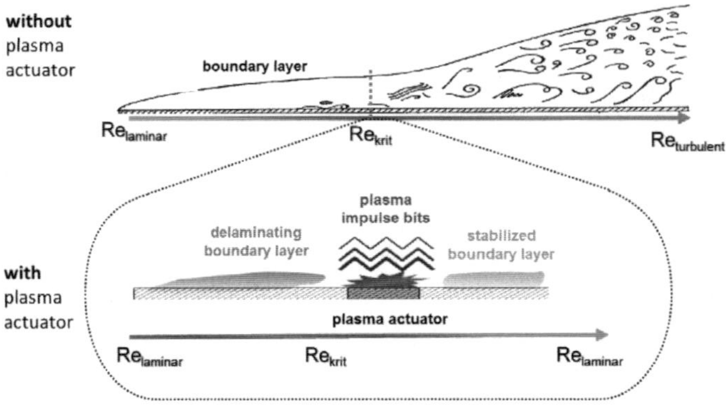

Figure 1: Schematics of the Tollmien Schlichting Wave creation (top) and function principle of plasma actuator for TSW counteraction (bottom).

consumption of this kind of actuator was found to be too high for the desired application. Using higher frequencies than 1 kHz with an inductive energy storage system as a power supply unit lead to a power consumption of more than 100 W per actuator.

Another type of plasma actuators, the so called spark jet-actuators, were also tested for flow control applications. However they are limited in their reachable maximum frequency (up to 5 kHz) and have as well a high power consumption. This frequency limitation results from the recovery times of the actuator after each pulse considering the gas viscosity [3]. The last and maybe best studied plasma actuator is based on a Dielectric Barrier Discharge (DBD). It can reach frequencies in the kHz regime with adjustable power input. However the frequency control and the required high voltages are the main problems for the use of this type of actuator [3].

For the application of plasma actuators for the drag reduction in LTPs several conditions must be fulfilled. They should be able to operate at pressures from 50 mBar to 100 mBar with an ignition voltage less then 1000 V to keep electrical circuit design complexity as low as possible. The actuator design and size should allow the integration into the LPT guiding blades. Moreover the surface of the actuators must not introduce a passive interaction with the flow. This means that the actuator surface needs to remain unchanged (no surface erosion or change due to heat) throughout its operation. The actuators also have to be controlled in frequency and moreover in energy output. For this reason only planar electrodes can achieve a minimum of flow disturbance.

Earlier studies have shown that glow discharges based actuators can influence flow and fulfill the requirements presented earlier [5]. Unlike DBD actuators the electrodes of the glow discharge actuator can be integrated any disturbance of the profile surface. It does not require such high voltages as needed for DBD actuator operation. This makes it easier to integrate into a given flow profile and electrical system.

In this work, a glow discharge plasma actuator is developed and tested for its applicability for TSW extinction in LPTs. The main aim is to create an actuator which is capable to operate at frequencies of up to 40 kHz without introducing significant thermal damage to the LPT surface.

2. Experimental Setup

In the actuator presented in this work the discharge is ignited between two cylindrical aluminum electrodes (⌀ 2 mm) with a center to center gap of 4.4 mm. The electrodes are embedded in a PEEK carrier plate. Additionally both electrodes are heat insulated with a ceramic tube against

IOP Conf. Series: Journal of Physics: Conf. Series **825** (2017) 012007 doi:10.1088/1742-6596/825/1/012007

the carrier plate. Schematic of the experimental setup is shown in fig. 2. A high voltage, high frequency power supply unit which consists of a simplified high voltage fly-back transformer circuit that is used to ignite a glow discharge between the two electrodes. The discharge is initiated by an external pulse generator of which its amplified signal is applied to a high speed switching MOSFET. The MOSFET is driven by a TC1427 optocoupled driver circuit allowing switching frequencies up the 4 MHz. The high voltage transformer can provide discharge voltages up to 40 kV with a free adjustable frequency up to 40 kHz due to power supply limitations. For the experiments frequencies of 1 kHz, 10 kHz and 20 kHz were used. The charging time of the transformer was set between 3.5 us to 5.5 us in order to keep thermal load on the MOSFET as low as possible.

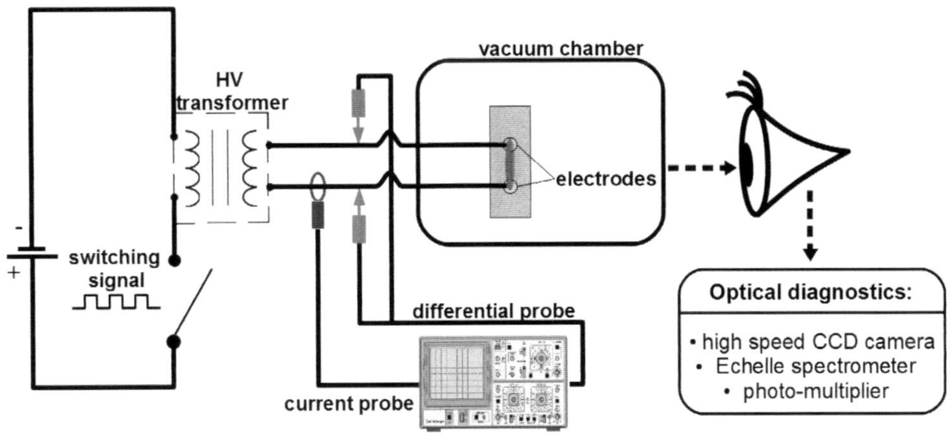

Figure 2: Schematic of the experimental setup.

The actuator was operated in a vacuum chamber under pressure values from 50 mbar to 100 mbar, which was previously filled with ambient air, in order to simulate pressure conditions in the LPT. At this pressure the electrodes work around the Paschen minimum for air. Thus, the breakdown voltage is significantly reduced from 3 kV to 350 V in normal air [6] [7].

For characterization of the discharge several diagnostics were used. Discharge voltage and current were monitored using a differential probe (ADP 305, Le Croy, Bandwidth: 20 MHz) and Pearson coil (model 310, 3dB cutoff at 10 MHz). A photo-multiplier (Type: 57X08 PM Section with pre-amplifier operated at 1.2kV) was used to temporally resolve the states of discharge. Furthermore emission plasma spectroscopy was applied to investigate plasma temperature and plasma composition. Here for an Echelle spectrometer (Aryelle 400, LTB Berlin, operation range from 200 nm to 900 nm, exposure time 60 s) with a spectral resolution $\frac{\lambda}{d\lambda} = 150000$ was used. The spatial distribution of the plasma was investigated using a high speed camera (PCO.dimax HD, 1 us exposure time).

3. Results

In fig. 3 current and voltage signals recorded at operation frequencies of 1 kHz, 10 kHz and 20 kHz are presented. The corresponding transformer charging duration was 5.5 µs for 1 kHz and 0.5 µs for 10 kHz and 10 kHz were applied. Furthermore the photo-multiplier signal registering the discharge emission and the discharge trigger signal were recorded.

The photo-multiplier signal confirms that the plasma is ignited for a definite time period. It can be observed that the ignition between both electrodes starts with the highest voltage peak around 6 µs after the MOSFET is switched off. The current through the plasma rises shortly

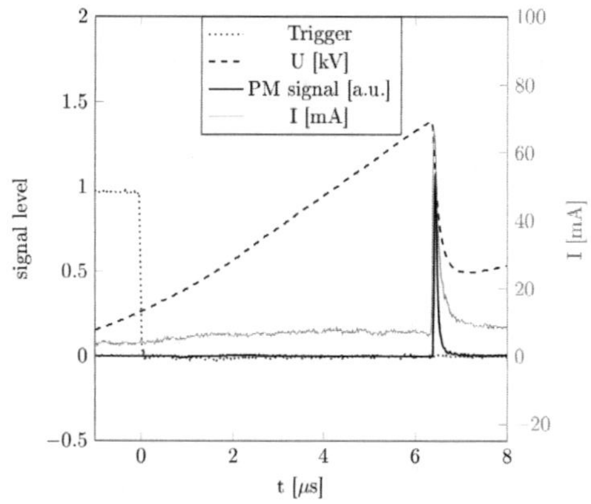

(a) 1 kHz

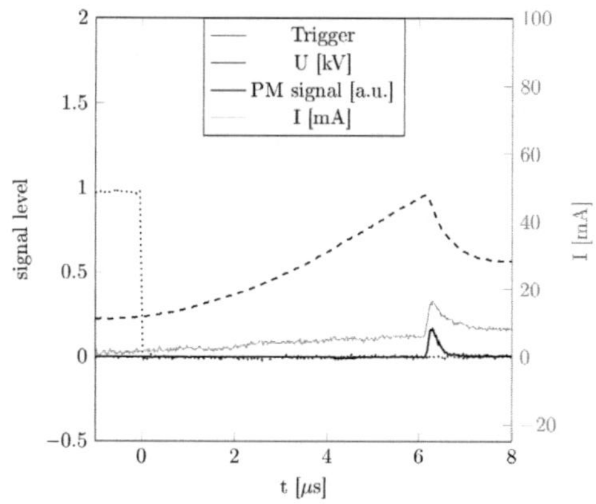

(b) 10 kHz

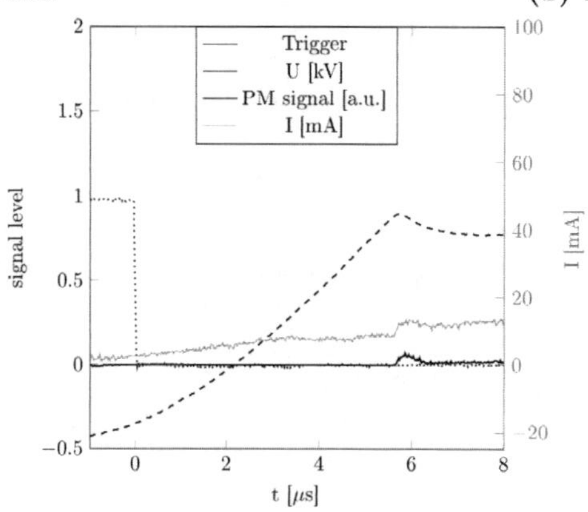

(c) 20 kHz

Figure 3: Electrical and optical characterization of the discharge for different discharge frequencies.

after the maximum voltage peak (ignition point), generating a glow discharge plasma, whose spectral emission is made visible with the photo-multiplier. It can be assumed that the plasma existence directly correlates with the photo-multiplier signal.

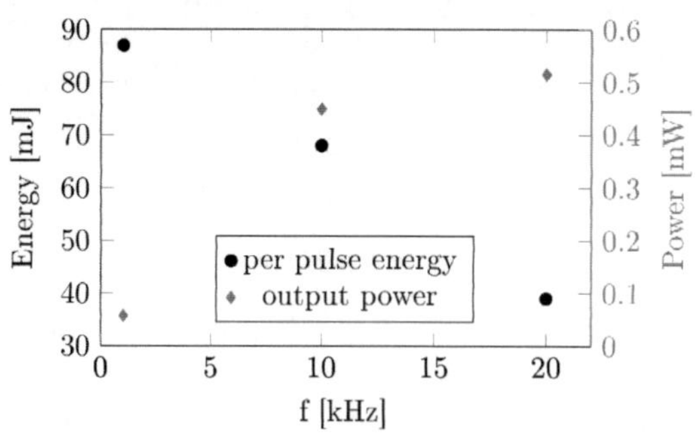

Figure 4: Plasma energy per pulse (black dots) compared to plasma power (red squares).

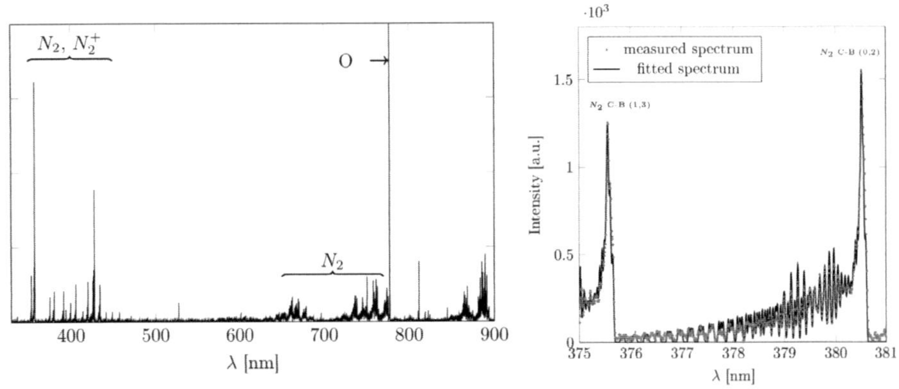

(a) overview spectrum

(b) second positive N_2 spectrum

Figure 5: Time and space averaged spectra used to obtain the overview of emitting species(a) and vibrational and rotational gas temperature determination using second positive system (b).

In all the measured current signals a nonzero current was observed before the actual plasma ignition point. It is most likely caused by a parasitic capacitance in the secondary transformer circuit. All voltage and current values are typical for a glow discharge. However the discharges vary in their maximum peak currents and voltages depending on the operation frequency. It can be stated, that both ignition voltage and discharge current decrease with increasing frequency. This can be explained by the limited power of the DC power supply used for the charging of the transformer. The overall discharge length of 0.6 µs was determined from the photo-multiplier signal as reference for all operation frequencies.

A plasma energy input calculation for each pulse was performed as shown in fig. 4. It is clearly visible, that an increase of the frequency leads to a decrease in the pulse energy but also to a significant increase of the power consumed by the plasma. This data shows that an optimization of the current actuator design is necessary to reduce the power consumption. This can be achieved by decreasing theinter-electrode distance; therefore lowering the breakdown voltage. Furthermore the actual PSU setup requires optimization. The oscillating effects show the frequency limitation of the actual setup, which are caused by higher losses in the electrical system.

In order to confirm that no metal vapor was present inside the plasma hence no electrode erosion was taking place during the discharge, emission spectroscopy measurements were performed. The resulting spectrum is shown in fig. 5. Only nitrogen and oxygen emission lines could be detected here. This leads to the assumption that the electrodes are not measurably eroded. The spectrum could also be used for the plasma temperature estimation using N_2 emission lines at 375.4 nm and 380.5 nm. Using the model described by Herzberg rotation temperatures 600 K to 700 K and vibration temperatures 1600 K to 5000 K were calculated [8].

The spatial distribution of the discharge were evaluated using high speed images. As shown in fig. 6 the cathode spot is clearly visible as the brightest part of the discharge. The light intensity changes to lower intensities when working at higher frequencies which correlates with the already mentioned less energetic pulses at higher frequencies. The side view clearly visualizes, that the glow discharge forms a bow line between the both electrodes. This shape remains constant with varying frequencies.

After testing the electrode setup for 30 minutes at 10 kHz no visible erosion of the electrodes or the carrier PEEK plate could be observed using a microscope.

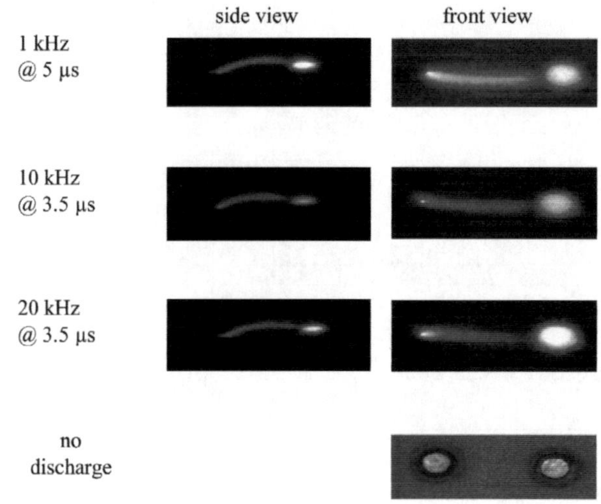

Figure 6: Images of the discharges taken at different discharge frequencies and corresponding charging times of the transformer.

4. Conclusion

In this work a concept for active flow control has been presented. It is based on a glow discharge with a self-made transformer circuit. Under operating conditions which were kept close to application in LPTs (pressure and frequencies) the actuator has shown a reliable and reproducible performance. It confirms, that the actual setup is suitable to be used for further investigations in the wind tunnel.

The next step will be a test with multiple actuators inside a wind tunnel with flow dynamics measurements to see if the actuator is capable of influencing Tollmien Schlichting waves. Furthermore the actuator power supply needs to be optimized for higher frequency range and for multiple actuators operation in order to deliver better power efficiency. For this optimized cooling of the MOSFETs and special designed on-board circuits could improve the switching characteristics. Moreover, an other high voltage transformer could improve the system.

5. Acknowledges

This research was supported by the University of the Federal Armed Forces of Germany.

References

[1] Niehuis R and Mack M 2015 *Active Flow and Combustion Control 2014* 322 ISSN 1860-0824 URL `http://dx.doi.org/10.1007/978-3-319-11967-0_1`

[2] Bayoda K D, Benard N and Moreau E 2015 *Journal of Applied Physics* **118** 063301 ISSN 1089-7550 URL `http://dx.doi.org/10.1063/1.4927844`

[3] Cattafesta L N and Sheplak M 2011 *Annual Review of Fluid Mechanics* **43** 247272 ISSN 1545-4479 URL `http://dx.doi.org/10.1146/annurev-fluid-122109-160634`

[4] Corke T C, Post M L and Orlov D M 2007 *Progress in Aerospace Sciences* **43** 193217 ISSN 0376-0421 URL `http://dx.doi.org/10.1016/j.paerosci.2007.06.001`

[5] Roth J R, Sherman D M and Wilkinson S P 2000 *AIAA Journal* **38** 11661172 ISSN 1533-385X URL `http://dx.doi.org/10.2514/2.1110`

[6] Kchler A 2005 *VDI-Buch* **2** 544 URL `http://dx.doi.org/10.1007/b138561`

[7] Paschen F 1889 *Annalen der Physik* **273** 6996 ISSN 1521-3889 URL http://dx.doi.org/10.1002/andp.18892730505
[8] Glockler G 1940 *The Journal of Physical Chemistry* **44** 682682 ISSN 1541-5740 URL http://dx.doi.org/10.1021/j150401a019

Innovation in the development of plasma propulsion devices in Israel

Dan R Lev

Electric Propulsion Researcher,
Space Propulsion Systems Dept., Rafael, Haifa 3102102, Israel

E-mail: Dan.R.Lev@gmail.com

Abstract. In this paper we review plasma propulsion development approach which focuses on innovation. We then bring the example of the state of Israel in general, and Rafael in particular, and show how it has adopted an innovative approach to develop a low power Hall thruster and a low current cathode. To present one special test-case of innovation we elaborate upon the development process of a heaterless hollow cathode that was developed at Rafael. In particular, by presenting the cathode characterization and wear test results we demonstrate that the heaterless cathode developed has a sufficiently wide operational range and may operate continuously for 1,500 hours without any measurable degradation in performance.

1. Introduction

Plasma propulsion is a particular plasma application in which gas is ionized and accelerated through applied electric and magnetic fields with the ultimate purpose of generating momentum for spacecraft maneuverability [1]. Using plasma propulsion over a hundred different spacecraft, mainly communication satellites, maintain their trajectories in space [2]and many more are expected to be launched in the upcoming decade [3] Unlike chemical propulsion, that harnesses its energy from the chemical bonds within the propellant; and therefore is energy limited, plasma propulsion is limited only by the electrical energy supply on-board the spacecraft. As such plasma propulsion can be used to accelerate propellants to velocities much higher than with chemical propulsion. Consequently, plasma propulsion is more propellant-efficient than chemical propulsion and sees an increasing use over the past few decades [4] In particular, the use of plasma propulsion for communication satellites has increased by approximately 800% between 1995 and 2011 (Figure 1) due to the significant propellant mass savings compared with conventional chemical propulsion [4]

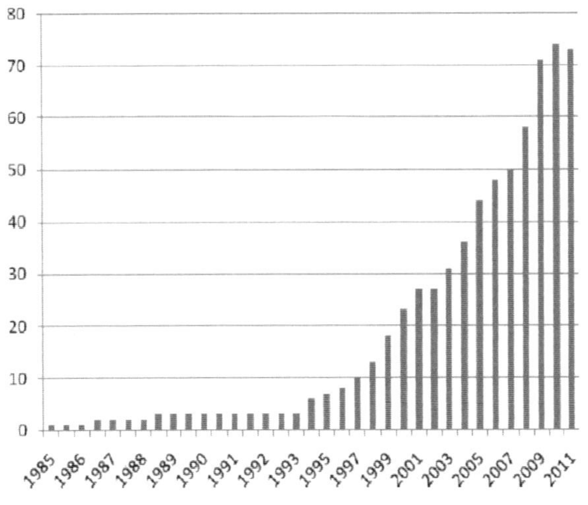

Figure 1. Number of operational communication satellites utilizing electric propulsion per year between the years 1985 and 2011 [4].

This ever-growing popularity of plasma propulsion drives many commercial entities to develop, manufacture and eventually offer plasma propulsion based systems for satellite platforms. However, due to the high capital and professional training requirements associated with the development of such systems these commercial entities have traditionally been subsidized by large financially-capable countries. As a consequence, only large geo-political entities, such as the US, EU, Russia, Japan and China, have produced full scale plasma propulsion capabilities which are being leveraged on to dominate the commercial plasma propulsion market. Nevertheless, one possible key approach taken by technologically oriented yet "ill-funded" entities to enter the plasma propulsion market is by offering innovative solutions for existing needs. Through innovation technological, thus commercial, advantage may be gained; therefore allowing new players to enter an existing plasma propulsion market.

In this paper we briefly discuss the innovation approach. We then bring two examples of innovative plasma propulsion devices developed at Rafael, Israel. Lastly, we focus and elaborate on one specific device by presenting a test-case of an innovative technical approach – the development of a heaterless hollow cathode that is conducted at Rafael.

2. Innovation approach

To succeed in penetrating an existing space market, which is largely dominated by large technologically-capable geo-political entities, small countries may seek to find innovative and alternative solutions to existing ones. In this approach the developer eliminates the notion of seeking conventional or traditional technical solutions, under the assumption that developing similar devices as already exist in the market will hinder future marketing efforts. Instead, the developer looks for new variations of traditional technological solutions, in combination with new technologies, which may out-perform existing solutions. Improvement on traditional technological configurations requires different thinking and innovation; whether from physics or engineering points of view. In this way small companies may gain technological advantages that outweigh the advantages of space-proven technologies.

The state of Israel, being a technologically-oriented yet a relatively ill funded geo-political entity, serves as a good example of a country practicing the innovative approach. We hereby present two particular examples of innovative plasma propulsion devices developed at Rafael, Israel. These are (1) CAM200 low power Hall thruster and (2) Rafael's heaterless hollow cathode neutralizer.

The developments of the two main components of Hall thruster based propulsion systems, namely the Hall thruster and the cathode neutralizer, include unconventional design that pose a physical advantage over the conventional configurations (Figure 2).

IOP Conf. Series: Journal of Physics: Conf. Series **825** (2017) 012008 doi:10.1088/1742-6596/825/1/012008

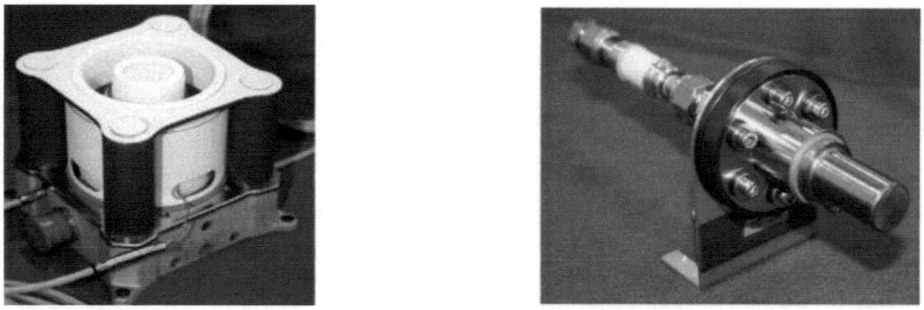

Figure 2. Picture of CAM200 low power Hall thruster (Left) and the heaterless hollow cathode (Right) both developed and manufactured at Rafael.

Israel's activities in the field of plasma propulsion began in the late 90's at Soreq where basic low power Hall thruster research was carried out at an academic level [5] During the first decade of this century all activities migrated to Rafael and the Technion [6]while building on the work and achievements gained a decade earlier. Since low power Hall thrusters already existed in the world at that time [7]a different thruster configuration was conceived, and later patented [8]to give the thruster better attributes compared with his low power Hall thruster cousins. The CAM200 Hall thruster, developed by Rafael, consists of co-axial anodes [9](Figure 3). The special anode configuration enables more efficient thruster operation than conventional Hall thrusters – making the thruster more attractive to potential customers.

Using theoretical, numerical and experimental investigations it was shown that the co-axial anode configuration modifies the electric field pattern within the thruster channel, acting to repel the ions from the thruster walls. This effect leads to ion beam focusing which gives a strong axial momentum component to the ejected ions. Ultimately, the well-directed ion beam carries a higher fraction of the input power to the thruster relative with other low power Hall thrusters; leading to higher efficiency. In addition, it was recently shown that CAM200 performance envelope is similar to high power Hall thrusters – the most efficient "members" of the Hall thruster device family [10].

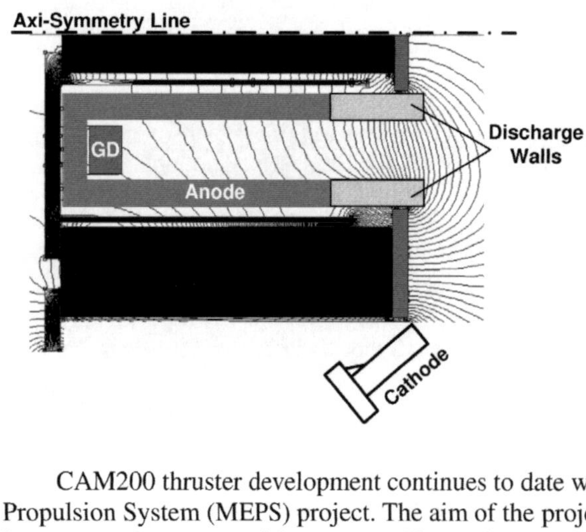

Figure 3. Two-dimensional configuration and magnetic field pattern of the CAM200 Hall thruster.

CAM200 thruster development continues to date within the frame of the Micro-satellite Electric Propulsion System (MEPS) project. The aim of the project is the development and qualification of a low power plasma propulsion system. The project is a joint development activity between Israel and Europe [11] In particular developers from Israel, Italy and Greece are in charge of the design, manufacturing and testing of the developed system. Within the frame of the MEPS project component responsibilities are distributed between the different parties, as well as financial responsibilities.

Future development phase shall include environmental tests, thruster lifetime experiment and plasma propulsion system integration tests.

The cathode neutralizer consists of a heaterless configuration that enables fast ignition along with higher reliability. In the next section we elaborate on the cathode development process.

3. Test-case for innovation – the heaterless hollow cathode

To further illustrate the 'innovation' approach mentioned above we present a test-case – the development of the heaterless hollow cathode that is conducted at Rafael. We first explain the different technological mechanisms of heaterless hollow cathodes compared with conventional cathodes. We then overview the development cycle of the heaterless hollow cathode and conclude by outlining the characterization and lifetime qualification tests of the cathode.

3.1. The role of the cathode neutralizer for plasma propulsion

The cathode neutralizer (Figure 4) is an electron-emitting device that serves three main functions: (1) supplying sufficient stream of electrons for the neutralization of the ion beam coming out of the thruster, (2) serving as the negative terminal of the electric propulsion circuit and (3) igniting the thruster-cathode module [12]

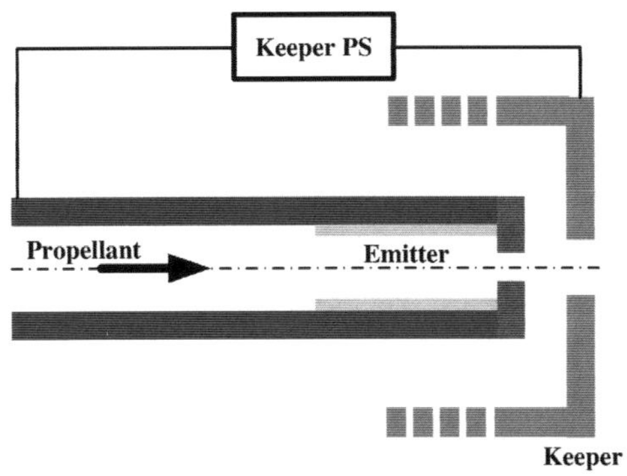

Figure 4. Basic configuration of a hollow cathode neutralizer for plasma propulsion applications.

The most important component within the cathode is the electron emitter, sometimes referred to as *insert* in literature. The emitter is made of a low work function material that emits electron flux in the order of ~1 A/cm^2 at relatively moderate temperatures of over 1,000°C [13]. Since for the thruster to operate the emitter has to be at this temperature both pre-heating of the emitter and proper cathode thermal design are required. The traditional emitter-heating method is by a dedicated heater element wrapped around the emitter tube. Prior to cathode ignition the heater is turned on and the emitter brought to its operational temperature and cathode ready to be turned on with the thruster. In many cathodes the thermal design is such that after thruster ignition sufficient heat flux from the cathode discharge maintains emitter temperature and the heater is no longer required.

In general, cathodes neutralizers should emit the amount of electron current required by the plasma thruster. As such cathodes tend to scale up or down with the thruster – low power Hall thrusters require low current cathodes.

3.2. Basic principles and advantages of heaterless hollow cathodes

Heaterless Hollow Cathodes (HHCs) are a subclass of hollow cathodes that do not require external heating to heat up the electron emitter to its operation temperature. Instead of using external heating HHCs use a unique ignition technique. Firstly, high voltage pulse is applied between the emitter and keeper (Figure 4) so to electrically breakdown the injected gas. Immediately after initial discharge creation a separate power supply controls the emitter-keeper current, a process during which the emitter is heated. The heating process lasts until the emitter reaches its operation temperature. Lastly, after steady discharge has been initiated the electric thruster is turned on by applying the required emitter-anode current. The entire thruster ignition duration is usually less than 100 seconds.

Since HHCs do not use external heating for ignition they possess advantages over their heater-utilizing cousins. First, In comparison to heater-utilizing cathodes that have typical readiness time in the order of minutes, due to the required heating duration, HHCs reach steady-state operation within tens of seconds. Secondly, since no external heating is required the Power Processing Unit (PPU) does not contain the corresponding heater module, therefore lowering its mass. Lastly, since cathode heaters usually contain refractory metals and experience extreme thermal cycling they are susceptible to thermal failure. HHC combat this problem by completely removing the ignition dependence on heaters.

HHCs have been explored for over four decades [13]in various locations around the globe. Most work was academic in nature and performed by either universities or research entities. Some work was presented by space industries or government agencies, yet little follow up literature exists from these attempts that most likely had no continuation. Nevertheless, in recent years there is a regrowing interest in HHC technology, an interest that might lead to maturation of the technology ending in space proven hardware.

3.3. The development cycle of the HHC at Rafael

Due to the technological advantages mentioned Rafael began its cathode development process by designing and investigating various ignition and operation techniques of HHCs. The ultimate purpose was to design a cathode able to supply discharge current of 0.5-1.1 A so to neutralize the ion beam ejected out from a low power Hall thruster. The development cycle (Figure 5) included the following phases:

(1) Design – in which the mechanical structure of the cathode was modelled with CAD methods.
(2) Simulations – in which the plasma discharge within the cathode was modelled and cathode's thermal and structural behaviours simulated.
(3) Manufacture – in which the cathode parts and components were manufactured and jointing processes developed.
(4) Experiments – in which the cathode hardware was experimented on in laboratory environment.
(5) Analysis – in which the experimental results and cathode post-operation structure were adequately analyzed and conclusions drawn.

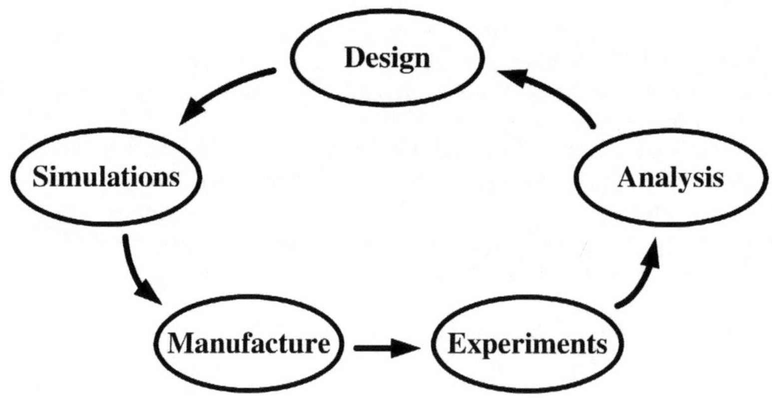

Figure 5. The development cycle of the heaterless hollow cathode as carried out at Rafael.

After the analysis phase conclusions drawn and lessons learned fed into the design phase and the cycle continued until desired cathode operation was achieved.

3.4. Cathode characterization and qualification

The development cycle converged into the most recent heaterless hollow cathode design (Figure 2). After verifying adequate and satisfactory cathode operation, two advanced operation verification phases were carried out – cathode function characterization and cathode 1,500 hr wear test [15]

During the characterization phase cathode operation was monitored for different operational parameter values; namely mass flow rate and discharge current. In particular, floating keeper-emitter voltage was monitored to assess cathode sheath variations with the operational parameters [16] The measured results are presented in Figure 6. The floating keeper-emitter voltage gives indication for the cathode "health level"; high voltage indicates on insufficient ion flux into the emitter surface and implies on possible cathode operation limitations. Additionally, large voltage fluctuations indicate on possible ion acoustic turbulence which consumes power from normal cathode operation in favour of an external ion energy increase [17]

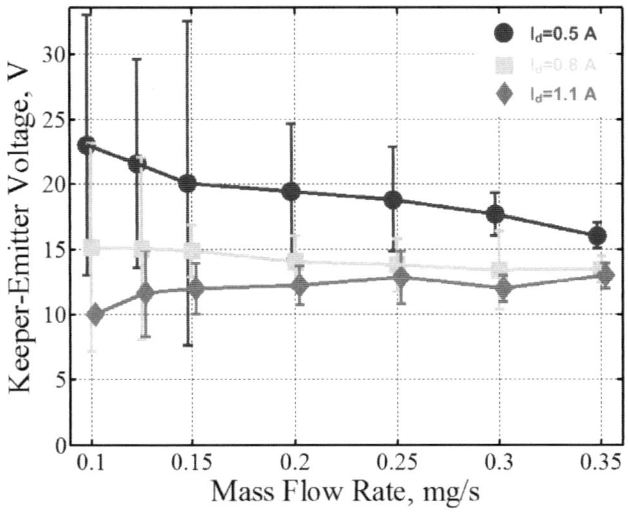

Figure 6. Floating keeper-emitter voltage plotted against cathode mass flow rate for different discharge current values.

The characterization process assures adequate cathode operation within the pre-defined limits and depicts cathode behaviour limits.

After the completion of the characterization process it was necessary to prove the ability to operate the cathode continuously throughout the projected lifetime of approximately 1,500 hours. For this reason a 1,500 hour cathode wear test was executed. During the wear test the cathode was operated continuously in a 24/7-manner while several electrical and thermal characteristics were tracked and recorded. In particular two cathode operation properties were monitored – (1) floating keeper-emitter voltage and (2) cathode keeper external temperature. The former gives indication on the electron emitter vitality and whether it reached its end-of-life. In case of emitter depletion its work function is expected to rise [18]leading to rise of cathode voltage, hence rise of floating keeper-emitter voltage [19] The latter also indicates on the electron emitter vitality as any increase in emitter work function leads to an increase in steady-state emitter temperature which will increase the temperatures throughout the cathode body.

Floating keeper-emitter voltage and cathode keeper temperature plotted against elapsed time over the course of 1,500 hours are presented in Figure 7. It can be observed from the figure that throughout the entire cathode wear test its operation is stable and does not deteriorate. We therefore conclude that the cathode has expected lifetime of over 1,500 hours.

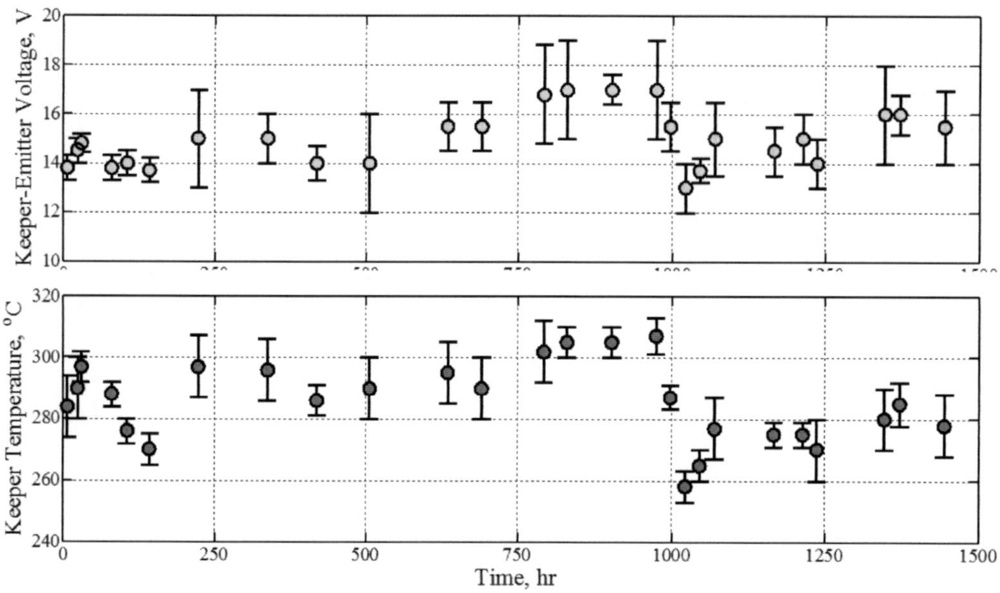

Figure 7. Floating keeper-emitter voltage (top) and cathode keeper temperature (bottom) plotted against elapsed time at the cathode nominal operational point (Discharge current: 0.8 A, Mass flow rate: 2.5 sccm of Xenon).

The successful completion of both the characterization and wear test phases gives the confidence that the developed product meets any expected requirements of such device. Using the development methodology described above the risk of innovation was mitigated and a new space commercial device was born.

4. Conclusions

In this paper we discussed the innovation approach for plasma propulsion device development used to produce cutting-edge solutions that outweigh the existing ones. We brought the example of two device developments carried out in Israel and showed the innovative technological features of each one. To present one special test-case of innovation we elaborated upon the development process of a heaterless hollow cathode that was developed at Rafael. In particular we showed the characterization and wear test phase experimental results. The data proves that the heaterless cathode developed has a sufficiently wide operational regime and may operate for 1,500 hours without any measurable degradation in performance.

References

[1] Jahn R and Choueiri E 2001 *Encyclopaedia of Physical Science and Technology, 3rd Ed.* **5**, 125-141

[2] Pollard J and Martinez-Sanchez M 1998. *J. Propulsion and Power* **14**(5), 688-699

[3] Euroconsult 2015 *Satellites to Be Built and Launched by 2024*, Technical report, Euroconsult

[4] Ganote S, Allen G, Dalby R and Wickham T 2013 *Satellite Electric Propulsion - Key Questions for Satellite Operators and Their Suppliers*. Technical report, Avascent

[5] Ashkenazy J, Raitses Y and Appelbaum G 1997 *IAEC Annual Report*, IL9906633

[6] Guelman M, Kapulkin A, Balabanov V, Rabinovich L and Appelbaum G 2008 A New Low Power Hall Thruster Concept, IAC-08-C4.4.6, *Proceedings of the 59th International Aeronautical Congress (IAC)*

[7] Ahedo E and Gallardo J 2003 Scaling Down Hall Thrusters, IEPC-03-0104, *Proceedings of the 29th International Electric Propulsion Conference (IEPC)*, Toulouse, France

[8] Kapulkin A Guelman M Balabanov V and Rubin B 2010 United States, Patent Application Publication, Pub. No.: US 2010/0107596 A1

[9] Lev D, Eytan R, Alon G, Warshavsky A and Appel L 2015 CAM200 Hall Thruster – Development Overview, IAC-15-C4.4.4, *Proceedings of the 66th International Aeronautical Congress (IAC)*, Jerusalem, Israel

[10] Lev D, Alon G, Mykytchuk D and Appel L 2016 Performance Characterization and Vibration Test of the CAM200 Low Power Hall Thruster, AIAA-2016-4831, *Proceedings of the 52nd Joint Propulsion Conference (JPC)* , Salt Lake City, UT, USA

[11] Misuri T, Albertoni R, Ducci C, Waldvogel B, Appel L, Eytan R, Lev D, Dannenmayer K and Di Cara D 2015 MEPS: A Low Power Electric Propulsion System for Small Satellites, IAA-B10-1103, *Proceedings of the 10th IAA Symposium on Small Satellites for Earth Observation*, Berlin, Germany

[12] Goebel D and Katz I 2006 *Fundamentals of Electric Propulsion: Ion and Hall Thrusters* (JPL Space Science and Technology Series)

[13] Polk J, Goebel D, Watkins R, Jameson K, Yoneshige L, Przybylowski J and Cho L 2006 Characterization of Hollow Cathode Performance and Thermal Behavior, AIAA_2005-4066, *Proceedings of the 42nd Joint Propulsion Conference (JPC)*, Sacramento, CA, USA

[14] Lev D and Appel L 2016 Heaterless Hollow Cathode Technology - A Critical Review, SP2016_3125366, *The 5th Space Propulsion Conference (SP)*, Rome, Italy

[15] Lev D, Alon G, Mykytchuk D and Appel L 2016 Characterization and Qualification of a Low Current Heaterless Hollow Cathode, AIAA-2016-4732, *Proceedings of the 52nd Joint Propulsion Conference (JPC)* , Salt Lake City, UT, USA

[16] Lev D, Alon G, Mykytchuk D and Appel L. 2015 Development of a Low Current Heaterless Hollow Cathode for Hall Thrusters, IEPC-2015-163, *Proceedings of the 34th International Electric Propulsion Conference (IEPC)*, Hyogo-Kobe, Japan

[17] Goebel D, Jameson K, Katz I and Mikellides I 2007 *Phys. Plasmas* **14** 103508

[18] Polk J, Mikellides I, Capece A and Katz I 2009 Barium Depletion in Hollow Cathode Emitters, AIAA-2009-5197, *Proceedings of the 45th Joint Propulsion Conference (JPC)* , Denver, CO, USA

[19] Cirri G, Cipriani A, Matticari G, Severi A, Bartoli C, Saccoccia G and Von Rohden H 1993 Review of Qualification Activities on the Neutralizer for the RIT10 Ion Thruster, IEPC-93-109, *Proceedings of the 23rd International Electric Propulsion Conference (IEPC)*, Seattle, WA, USA

Pulsed arc plasma jet synchronized with drop-on-demand dispenser

F Mavier[1], L Lemesre, V Rat, M Bienia, M Lejeune and J-F Coudert

[1] Univ. Limoges, CNRS, ENSCI, SPCTS, UMR 7315, F-87000 Limoges, France.

fabrice.mavier@etu.unilim.fr

Abstract. This work concerns with the liquid injection in arc plasma spraying for the development of finely structured ceramics coatings. Nanostructured coatings can be now achieved with nanopowders dispersed in a liquid (SPS: Suspension Plasma Spraying) or with a salt dissolved into a liquid (SPPS: Solution Precursor Plasma Spraying) injected into the plasma jet. Controlling electric arc instabilities confined in non-transferred arc plasma torch is therefore a key issue to get reproducible coating properties. Adjustment of parameters with a mono-cathode arc plasma allows a new resonance mode called "Mosquito". A pulsed arc plasma producing a periodic regular voltage signal with modulation of enthalpy is obtained. The basic idea is to synchronize the injection system with the arc to introduce the liquid material in each plasma oscillation in the same conditions, in order to control the plasma treatment of the material in-fly. A custom-developed pulsed arc plasma torch is used with a drop-on-demand dispenser triggered by the arc voltage. A delay is added to adjust the droplets emission time and their penetration into the plasma gusts. Indeed, the treatment of droplets is also shown to be dependent on this injection delay. A TiO_2 suspension and an aqueous solution of aluminium nitrate were optimized to get ejectable inks forming individual droplets. The feasibility of the process was demonstrated for SPS and SPPS techniques. Coatings from the suspension and the solution were achieved. First synchronized sprayings show a good penetration of the droplets into the plasma. Coatings show a fine structure of cauliflowers shapes. The synchronization of the ejection allows a control of morphology and a better deposition efficiency. Further investigations will find the optimal operating parameters to show the full potential of this original liquid injection technique.

1. Introduction

In the field of thermal spray coating processes, research has led to the development of nanostructured coatings by suspension plasma spraying (SPS) and precursor solution plasma spraying (SPPS). Liquid injections are promising techniques with the potential to become industrially viable [1,2]. However, a better control of plasma/material interactions is necessary. Mono-electrode DC plasma torches indeed generate strongly fluctuating plasma that modifies the thermal and dynamic transfers to the injected suspension droplets, resulting in an inhomogeneous treatment of the latter. This directly influences the texture and microstructure of deposits and subsequently their properties [3,4].

Efforts to understand the origins of these instabilities have been made. Previous works have shown that these instabilities are mainly due to the effects of compressibility of the plasma gas in the cathode cavity effects, belonging to the instability mode called Helmholtz mode. Other fluctuations are due to successive phenomena of elongation, breakdown and restrike of the electric arc, also called "restrike mode". In alternative to instabilities attenuations, a new approach is proposed: the reinforcement and modulation of the instabilities [5]. The adjustment of process parameters has allowed obtaining a pulsed laminar plasma with a modulation of its properties. A low powered DC torch is used and operates with pure nitrogen as plasma forming gas. This device is synchronized with a drop-on-demand injection system to reproduce the same conditions of plasma/material interactions for each injected droplet [6]. Aluminum nitrate aqueous solutions and TiO_2 suspensions are injected to make homogeneous coatings with a controlled microstructure and chemical composition. The objectives of this work are (i) to characterize and understand plasma / droplet heat and dynamics transfers and (ii) to highlight the influence of the synchronization on the coatings obtained.

2. Experimental setup

2.1. The pulsed arc plasma

The spray depositions in this study are carried out using an experimental DC torch, custom-developed at the University of Limoges, working in the "Mosquito" pulsed mode. In figure 1 we can see a pulsed arc plasma and a modulated arc voltage signal. CCD camera shows individual plasma gusts emitted at the same frequency as arc voltage oscillations. The plasma gas is nitrogen. The current is 15 A and arc voltage oscillates between 40 and 100 V at ~1400 Hz. The average specific enthalpy obtained by performing a calorimetric measurement on the water cooling system of the torch is 13 $MJ.kg^{-1}$. The operating torch parameters are shown in table 1.

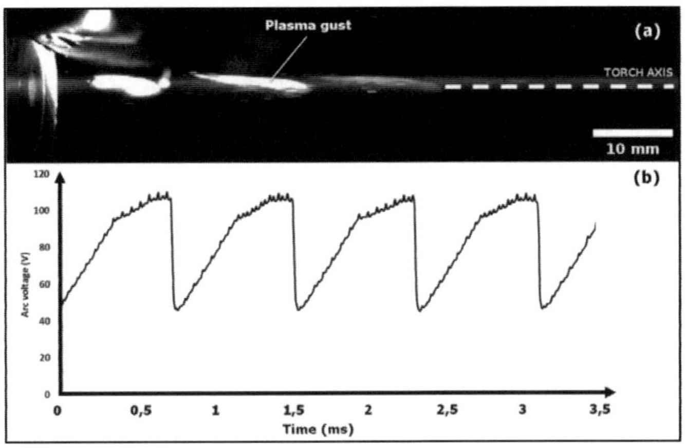

Figure 1. Imaging of a plasma jet in the "Mosquito" pulsed mode, camera aperture time: 60μs (a), arc voltage signal in the pulsed mode (b).

Table 1. Plasma spraying operating parameters.

Parameter	Value
Current	15 A
Average enthalpy	13 $MJ.kg^{-1}$
Anode nozzle diameter	4 mm
Plasma gas	Nitrogen
Flow rate	2slm

2.2. The drop-on-demand (DOD) dispenser

A drop-on-demand dispenser is used to inject the liquid material (a suspension or a solution). The dispenser nozzle where the droplets are emitted is fixed on the top (around 10 mm) of the plasma torch nozzle exit (figure 2). The diameter of the droplets is 80 μm. The DOD dispenser is a glass capillary. The ejection is actuated by a piezoelectric crystal to generate an individual droplet for each pulse trigger. For this study, aluminum nitrate aqueous solutions and TiO_2 suspensions were used. The suspension is a 15 vol% commercial aqueous suspension of rutile with submicronic particles with a bimodal distribution of 70 and 350 nm (Ceradrop, France). The aluminum nitrate solution is custom-made according to a method which is described in the results section.

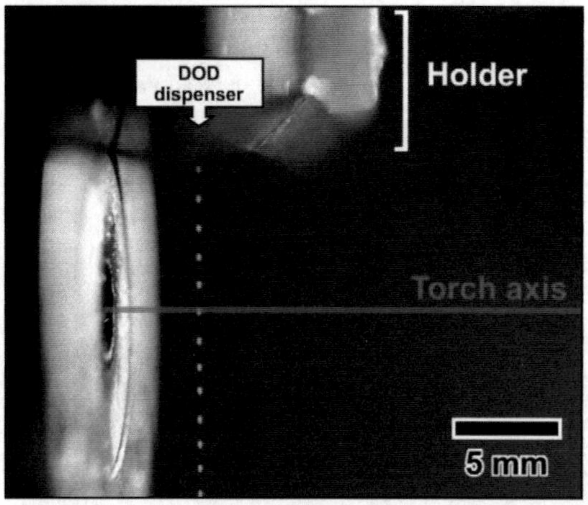

Figure 2. Picture of optimized ink ejection test at 1400 Hz without plasma, standard camera no synchronized, aperture time: 125μs.

2.3. Synchronization setup

The torch voltage is sampled by means of a dividing bridge and sent to an electronic synchronization device able to generate a set of delayed TTL pulses that trigger the drop-on-demand dispenser and the instruments that are used for the in-flight diagnosis (figure 3).

In-flight interactions are recorded by a CCD camera, coupled with a 50 W laser diode to illuminate the droplets at the emission wavelength of 801 nm. Optical emission spectroscopy (OES) is performed by using an Isoplane spectrograph (Princeton Instruments, Trenton, USA) associated with a high resolution Intensified CCD camera (PIMAX4 1024i, (1024 x 1024 pixels, 12.6 μm pixel size, Princeton Instruments, Trenton, USA). The camera, laser and OES are triggered by the synchronization system. A standard camera (Nikon D7100), not synchronized, is used for imaging.

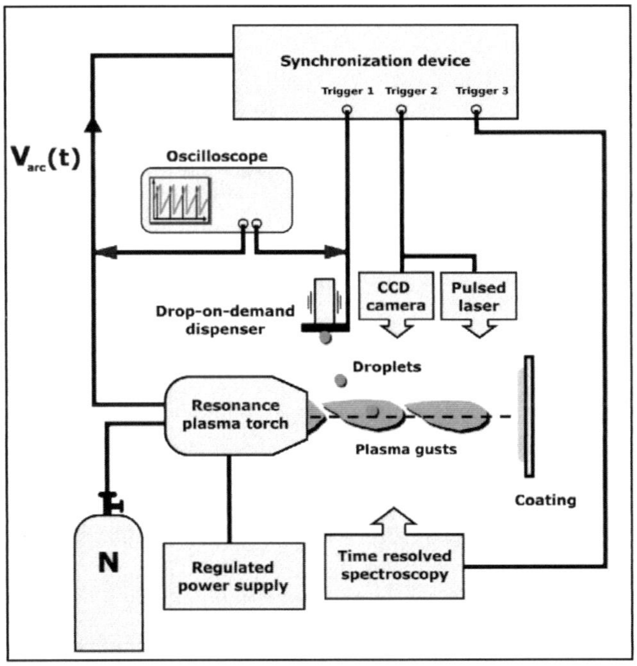

Figure 3. Schematic view of the synchronous injection system "Mosquitorch".

The spraying distance is fixed to 33 mm from the torch nozzle exit. This distance has given the best results in previous tests. Substrates are 25×25×1 mm mirror-polished stainless steel squares. The effective deposition time is fixed to 1 min to limit excessive thermal expansion stresses.

2.4. Coatings characterization

Coating morphologies and elemental composition are characterized on a scanning electron microscope (SEM, XL30, Philips, The Netherlands) coupled with an electron dispersion spectrometer (EDS). Imaging and diffraction pattern with transmission electron microscopy (TEM, JEOL2010, UK) are also used to characterize crystallinity and crystallite sizes. The structural characteristics of the coatings are determined using X-ray diffraction (XRD) with Cu Kα radiation (Siemens D 5000, Germany). The resulting diffractograms are indexed with JCPDS-ICDD database.

3. Results and discussion

3.1. Ink optimization and ejection tests

The propagation velocity of the acoustic wave within the glass capillary of the DOD microdispenser is estimated with this equation [7]:

$$a = \frac{c}{\sqrt{1+\frac{KdC1}{Eh}}} \quad (1)$$

where c is the speed of sound in the fluid (m.s^{-1}); $K=c^2\rho$ (GPa) ; C_1 a correction factor ; d the tube diameter (µm) ; h the thickness of the wall (µm) ; E glass Poisson's ratio.

This velocity is used to estimate an order of magnitude of the time taken by the wave between the excitation of the piezoelectric component and the ejection of a droplet to the orifice. This time is estimated at 18 µs, less than a Mosquito period of 700 µs. Therefore, the operating frequency of the

Mosquito is compatible with the DOD dispenser as the generation of two consecutive droplets does not cause disturbance.

The solution is prepared from aluminum nitrate salt 27 wt% ($Al(NO_3)_3.9H_2O$, 98%, Alfa Aesar, Germany) dissolved into deionized water. The suspension and the solution are optimized to fit the viscosity and surface tension requirements of the DOD dispenser given by the manufacturer. In addition, the ejection ratio Z, which involves Reynolds (Re) and Weber (We) numbers, must be in a range of 1 – 10 (equation 4) [8].

$$Re = \frac{\rho\, v\, r}{\eta} \quad (2) \qquad We = \frac{\rho\, v^2\, r}{\gamma} \quad (3)$$

$$Z = \frac{Re}{\sqrt{We}} = \frac{\sqrt{\rho\, \gamma\, r}}{\eta} \quad (4)$$

where ρ is the droplet density ($kg.m^{-3}$); γ the surface tension ($N.m^{-1}$); r the radius of the droplet (μm); η the viscosity (Pa.s).

Surface tension of fluids is controlled by a tensiometer (DCAT11, DataPhysics Instruments, Germany) and viscosity by a rheometer (ARG2, TA Instruments, USA). The commercial suspension has already been optimized for Z=3.3 and was used as purchased. In the case of the solution, glycerol (99+%, Alfa Aesar, Germany) is added to increase the viscosity of the initial solution (1.7 to 5.1 mPa.s) and a surfactant is added to decrease the surface tension (76,5 to 41,6 $mN.m^{-1}$) which leads to a ratio of ejection Z=8.8.

After these adjustments, the actuation pulse waveform sent to the piezoelectric printhead is optimized in order to get a stable ejection. An ejection is considered stable if a single spherical droplet is obtained without filaments or satellite droplets and whose velocity and trajectory are the same for all ejected droplets. In addition, preliminary studies showed that the droplets velocity must be higher than 2 $m.s^{-1}$ to obtain a good penetration of the droplets inside the plasma gusts emitted by the pulsed arc torch as shown in figure 1. For this study, the pulse is optimized to eject droplets at 2.5 $m.s^{-1}$. A cooling system with a copper holder and a cold water flow is used to prevent a temperature elevation of the DOD dispenser. Indeed, the temperature fluctuation strongly impacts viscosity which is the predominant term in the ejection ratio.

3.2. TiO_2 by SPS: In-flight diagnosis

Compared to conventional injection, e.g. mechanical injection, which delivers a continuous stream, drop-on-demand (DOD) systems allow isolating the droplets injection with the same experimental behavior thanks to the synchronized emission and acquisition. A regular jet of individual droplets is emitted from the DOD dispenser above the plasma jet, at the same frequency as the plasma gusts. In the first moments after the ejection, the droplet meets a gust of plasma resulting from the resonance mode. The temperature was measured between 6500 and 8000 K depending on the injection delay [9]. When a droplet arrives on the torch axis, it is fragmented in all directions, even at counter flow (figure 4). We suggest that this step is a thermal fragmentation because of low plasma velocity (~66 $m.s^{-1}$) and a Weber number below 0.5 (0.04-0.48). Indeed, mechanical fragmentation occurs when We > 12-14. Acceleration time of a droplet in the plasma is estimated at 800 μs while vaporization time is estimated at 200 μs, which implies that vaporization takes place more rapidly than acceleration.

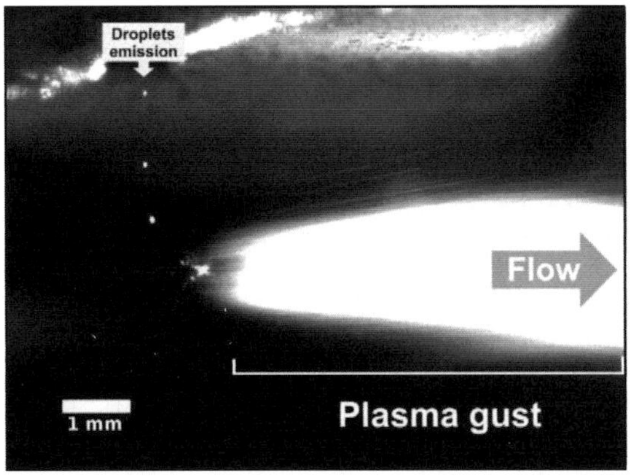

Figure 4. Synchronous injection of TiO_2 suspension in the pulsed plasma. Time-resolved imaging synchronized with droplet emission driven at the torch voltage frequency (laser shot: 20μs).

After fragmentation and vaporization, the droplet vapor seeds the plasma gust. Visually (figure 5) near the injection point blue light emission from plasma caused by the injected material is observed, which was attributed to titanium vapor by emission spectroscopy analysis. Titanium vapor is observed because heat transfer is efficient with nitrogen as plasma forming gas and the first excited states of titanium atoms are relatively low (<4eV). At larger distance, plasma gusts become orange corresponding to titanium oxide emission (TiO). Indeed, titanium oxide molecular bands are observed at 12mm from the nozzle exit of the plasma jet that could suggest that TiO forms with the in-flight oxidation of titanium atoms (figure 5).

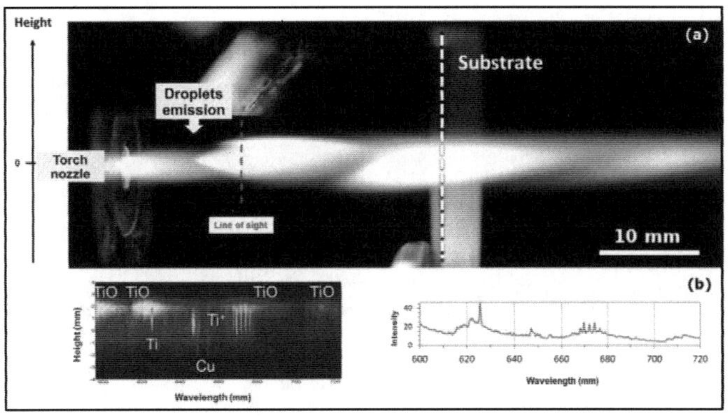

Figure 5. (a) Picture recorded with a standard camera; aperture time: 125 μs, no triggering. (b) Spectral mapping of light sampled at 12 mm from the torch exit following the line of sight.

3.3. TiO_2 by SPS: Coatings characterization

Two sets of TiO_2 coatings are investigated: coatings with and without synchronization injection. Coatings obtained without synchronized injection means that each injected droplet undergoes a different treatment and the injection is not mastered. They show a large panel of particles: submicronic fine structures of cauliflowers shapes, splats, molten particles and spherical particles from 10 micrometers (figure 6). XRD indexation shows only rutile phase (JCPDS No. 04-016-2838) for coatings with synchronized and non-synchronized injection.

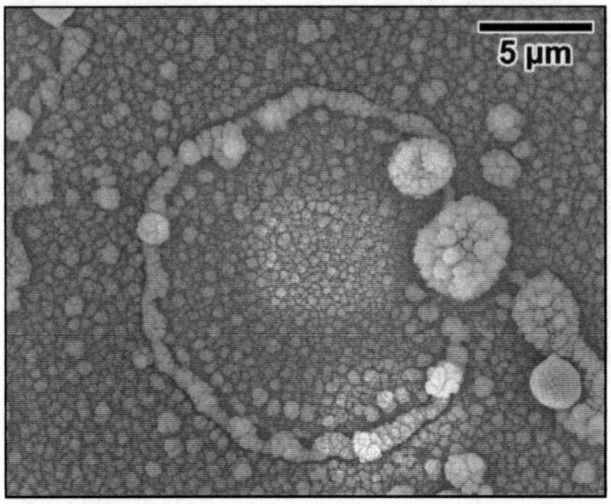

Figure 6. Synchronous injection of TiO_2 suspension in the pulsed plasma. Time-resolved imaging synchronized with droplet emission driven at the torch voltage frequency (laser shot: 20µs).

Coatings with a synchronized injection show a much better homogeneity, a large majority of the surface is composed with cauliflowers shapes (figure 7) suggesting the synchronization allows a control of the treatment and ultimately the morphology of the coating.

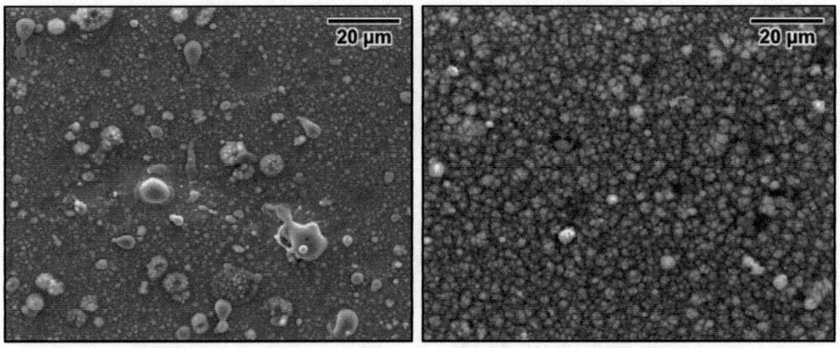

Figure 7. SEM picture × 750 top surface coating of TiO_2 with non-synchronized injection (left) and synchronized injection (right).

The thickness of the deposits is 10 microns. Cross-sections show a very porous columnar base (figure 8). High magnification shows grains from few tens of nm and a finer porosity in each cauliflower structure. TEM diffraction patterns confirm the presence of rutile, no amorphous phases are visible. Two crystallite sizes are observed: large grains of tens of nanometers [30-80 nm] and small grains <10 nm (figure 9). This high specific surface could be interesting for photocatalytic applications.

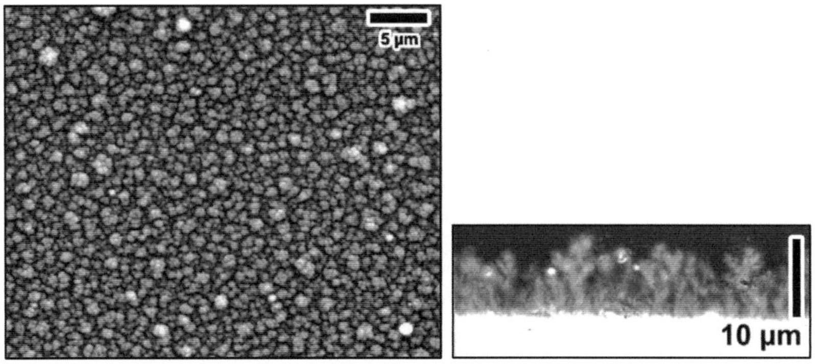

Figure 8. SEM picture ×2000 top surface (left) and a cross section (right) with synchronized injection.

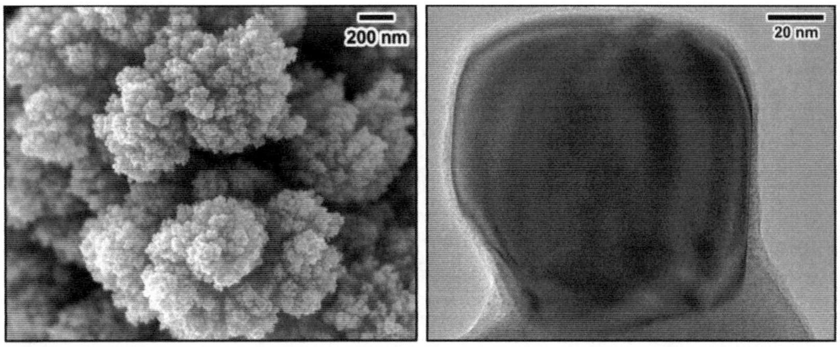

Figure 9. SEM picture top surface ×50 000 on a cauliflower shape (left) and TEM picture high magnification of a crystallite (right).

3.4. AlN by SPPS: Coatings characterization

Solution sprayings permits to obtain a fine structure similar to previous results by SPS. XRD confirms the presence of cubic aluminum nitride only (JCPDS No. 04-004-8344). The formation of nitride could be due to the use of nitrogen as plasma gas and the nitrate as precursor. First coatings obtain without synchronization show a structure less heterogeneous in comparison to the previous study by SPS. Some micrometrics splats and voids are observed (figure 10). Coatings with synchronized injection show a better homogeneity with no splats and no voids, the fine structure of cauliflowers shapes is the majority of the coating (figure 11). In addition, coatings with synchronized injection show a significant reduction in deposited material losses in the periphery of the substrate due to poor penetration into the plasma jet. The coatings have larger thickness: 9 microns for synchronized injection against about 5 microns to non-synchronized, with a much more even thickness distribution. The synchronizing effect on the morphology of the deposit is less impressive than for SPS tests, but it allows much more morphology control and a greater deposition efficiency.

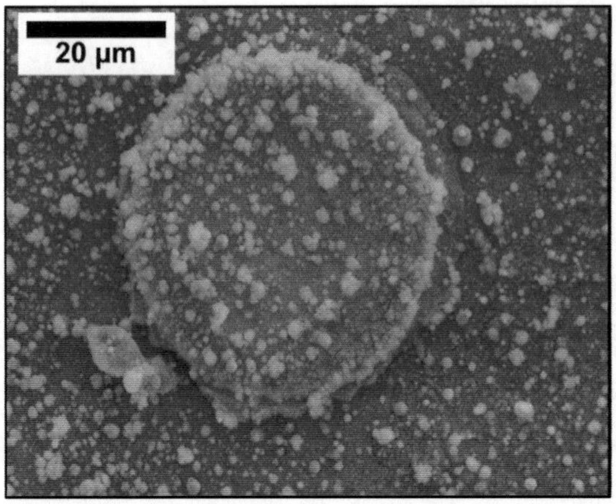

Figure 10. SEM picture top surface ×1200 SPPS coatings by non-synchronized injection.

Figure 11. SEM picture top surface ×120 SPPS coatings by non-synchronized injection (left) and synchronized injection (right).

A larger magnification on coatings with synchronized injection shows a fine structure of cauliflowers shapes from ~1 μm similar to TiO_2 by SPS, probably due to an aluminum vapor condensation on substrate and spherical structures of a few micrometers, perhaps originating from a nucleation in-flight process (figure 12). Candidato and al. show similar cauliflowers shapes by SPPS method, only with high concentration of dissolved precursors. They proposed that these microstructures are formed in-flight by a nucleation phenomenon [10]. Our future investigations will identify the formation process of these structures and optimize the operating parameters, e.g. spraying distance, to see if other morphologies and crystalline structures are possible via the synchronous injection process.

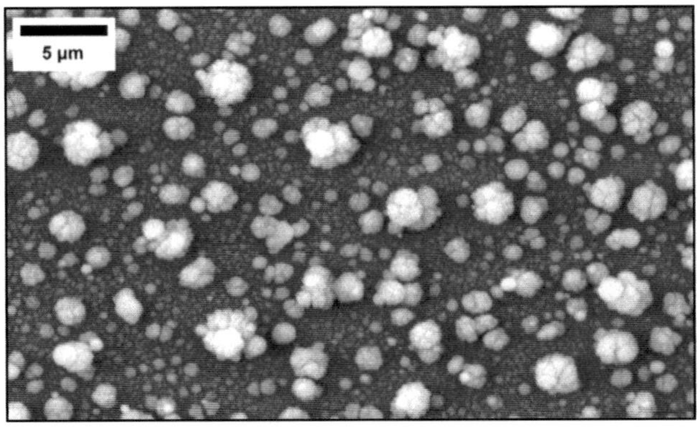

Figure 12. SEM picture top surface ×3000 SPPS coatings by synchronized injection.

4. Conclusion

A process combining atmospheric plasma spraying and inkjet drop-on-demand injection was developed. A pulsed arc plasma is generated with an arc plasma torch custom-developed at the University of Limoges. The principle is to inject each droplet in each plasma oscillation to control the dynamic and the enthalpy transfer to the material in order to increase the quality of the coatings. Inks have been optimized for injecting individual droplets at the same frequency as the plasma on the axis of the torch. In-flight diagnosis shows a good penetration of droplets in each plasma gust. The droplet travels to the torch axis and seeded the plasma gust creating a vapor by thermal fragmentation. First coatings with synchronized injection were made with an aqueous nano-suspension of TiO_2 and an aqueous aluminum nitrate solution. The results show a fine structure of cauliflowers shapes of a few micrometers in width and grains sizes of a few nanometers. Synchronization increases the homogeneity and the deposition efficiency of the coatings. Futures studies should be investigating the photocatalytic activity of our coatings. The use of an anatase suspension and adjusting operating parameters to maintain the crystalline species should be considered. Investigations will be identifying the optimizing parameters to show the full potential of this original process.

References

[1] Fauchais P, Vardelle M, Goutier S and Vardelle A 2015 Key Challenges and Opportunities in Suspension and Solution Plasma Spraying *Chem. Plasma Process* **35** 511–525
[2] Sampath S 2010 Thermal Spray Applications in Electronics and Sensors: Past, Present, and Future *Journal of Thermal Spray Technology* **19** 921–949
[3] Etchart-Salas R 2007 Suspension Plasma Spraying. An-alytical and experimental approach of the phenomena imply in the reproducibility and the quality of the deposits *Thesis (University of Limoges)*
[4] Etchart-Salas R, Rat V, Coudert J F, Fauchais P, Caron N, Wittman K and Alexandre S 2007 Influence of plasma instabilities in ceramic suspension plasma spraying *J. Thermal Spray Technol.* **16** 857–865
[5] Rat V and Coudert J F 2010 Pressure and Arc Voltage Coupling in Dc Plasma Torches: Identification and Extraction of Oscillation Modes *Journal of Applied Physics* **108** (4)
[6] Krowka J, Rat V and Coudert J F 2013 Resonant Mode for a Dc Plasma Spray Torch by Means of Pressure–voltage Coupling: Application to Synchro-nized Liquid Injection *Journal of Physics D: Applied Physics* **46** (22)
[7] Fromm J 1984 *IBM J. Res. Dev.* **28** 322–333
[8] Ainsley C, Reis N and Derby B 2002 Freeform fabrication by controlled droplet deposition of

powder filled mels *Journal of materials science* **37** 3155–3161
[9] Rat V, Krowka J and Coudert J F 2015 Modulation of the specific enthalpy of a pulsed arc plasma jet *Plasma Sources Sci. Technol.* **24** 045009
[10] Candidato Jr. R T, Sokołowski P, Nana G L, Constantinescu C, Denoirjean A and Pawłowski L 2016 Possibility of Developing Hydroxyapatite Coatings with High Purity by Solution Precursor Plasma Spray Process *(To be published)*

Acknowledgments
The French National Research Agency is thanked for financial support in the frame of PLASMAT program (ANR-12-JCJCJS09-0006-01). The Electric Arc Association (AAE) is thanked for financial support.

Optical emission spectroscopic study of Ar/H_2/CH_4 plasma during the production of graphene nano-flakes by induction plasma synthesis

Antaryami Mohanta, Briac Lanfant, Mehari Asfaha and Marc Leparoux
EMPA–Swiss Federal Laboratories for Materials Science and Technology, Laboratory for Advanced Materials Processing, Feuerwerkerstrasse 39, 3602 Thun, Switzerland

Email: antaryami.mohanta@empa.ch

Abstract. Graphene nano-flakes using CH_4 precursor were synthesized in a radio frequency inductively coupled plasma reactor with in-situ investigation of Ar/H_2/CH_4 plasma by optical emission spectroscopy at fixed H_2 and Ar flow rates of 4 and 75 slpm, respectively, and at different plate powers (12 to 18 kW), pressures (400 to 700 mbar) and CH_4 flow rates (0.3 to 2 slpm). Emissions from C_2 Swan band, C_3, CH and H_2 are observed in the optical emission spectra of Ar/H_2/CH_4 plasma. Plasma temperature estimated analyzing the C_2 Swan band emission intensities is found to be decreased with increasing pressure and decreasing plate power. The decreasing plasma temperature gives rise to increase in production rate due to increase in condensation process. The production rate is observed to be increased from 0 to 0.3 g/h at 18 kW and from 0 to 1 g/h at 15 kW with increase in pressure from 400 to 700 mbar at fixed CH_4 flow rate of 0.7 slpm. Broad band continuum emission appears in the emission spectra at specific growth conditions in which the formation of vapor phase nanoparticles due to condensation of supersaturated vapor is facilitated. The production rate at 12 kW, 700 mbar, and 0.7 slpm of CH_4 flow rate is found to be 1.7 g/h which is more than that at 15 and 18 kW. Thus, the broadband continuum emission dominates the optical emission spectra at 12 kW due to lower temperature and higher production rate, and is attributed to the emission from suspended nanoparticles formed in vapor phase. The synthesized nanoparticles exhibit flake like structures having average length and width about 200 and 100 nm, respectively, irrespective of the growth conditions. Nano-flakes have thickness between 3.7 to 7.5 nm and are composed of 11 to 22 graphene layers depending on the growth conditions. The intensity ratio (I_D/I_G) of D and G band observed in the Raman spectra is less than 0.33 which indicates good quality of the synthesized graphene nano-flakes.

1. Introduction

Synthesis of nanoparticles has been gained significant attention due to remarkably different physical and chemical properties of nanoparticles from their bulk counterpart in spite of possessing the same constituents [1, 2]. Many different techniques have been employed to synthesize various nanoparticles for different applications. The commonly used synthesis methods are colloidal chemistry, sol-gel methods, laser ablation and photochemical synthesis [3, 4]. Nanoparticles synthesized using chemical routes are not contaminate free due to the presence of other ions and reaction products from the reducing reactions [4]. Laser ablation synthesis process produces contaminant free high purity nanoparticles since nanoparticles are formed due to condensation of laser ablated plasma from a high purity target material and it does not involve chemical reagents in solutions [3, 4]. However, laser ablation technique cannot produce high quantity of nanoparticles at low cost for industrial applications. Radio frequency (RF) inductively coupled plasma (ICP) synthesis technique is an alternative plasma route for the synthesis of nanoparticles which is better than conventional plasma due to a large volume, high-temperature plasma zone, operating under different atmospheres, permitting synthesis of a great variety of nanoparticles, higher production rate with lower processing cost [5]. Furthermore, electrodeless operation of the ICP torch in the inert gas atmosphere involves clean plasma that ensures synthesis of high purity products [6]. Thus, RF thermal plasma is suitable for industrial scale mass production of high purity materials as it is a continuous process to accomplish high production volume with good synthesis control from the gaseous phase. A wide range of nanoparticles has been produced using appropriate precursors in RF inductively coupled plasma reactor [7-10]. In RF ICP synthesis system, the injected precursors convert into plasma which comprises mainly of electrons, ions and neutrals, and are the main source of the nanoparticles. Thus, it is imperative to characterize the plasma of the precursor during the synthesis of nanoparticles. Optical emission spectroscopy (OES) is a sensitive and non-evasive in-situ investigation technique to probe different constituents of plasma which provides useful information on different excited states in the plasma [11]. A detailed analysis of the emission lines observed at different spectral positions in the optical emission spectra reveals some important characteristics of the investigated plasma. The instrumentation associated with the OES study of plasma is very simple and straightforward which mainly involves simple optics to collect light emission from the plasma and a spectrometer to record the emission spectra. Plasma produced by several methods has been characterized by OES for different applications [12-15]. Ma et al. [12] have conducted OES measurements to investigate the microwave activated $Ar/H_2/CH_4$ plasma during the chemical vapor deposition of polycrystalline diamond. They have monitored the emissions from electronically excited H and Ar atoms, and C_2 and CH radicals as function of Ar and CH_4 flow rates, input power and pressure, and proposed that different species have different formation mechanisms since they exhibit different behaviors. Zhou et al. [13] have performed OES study and reported the existence of abundant C_2 and CH species in CH_4/Ar plasma along with C_3 species and/or 4-carbon clusters. They attributed CH radicals to be responsible for the trans-polyacetylene composition in the grain boundaries of deposited nano-crystalline diamond film. In recent years, synthesis of high quality graphene structures has been received increasing attention due to its distinct properties and potential applications in nanoelectronics [16]. Several studies have been reported on the synthesis of graphene nano-flakes with theoretical modelling based on fundamental theory and CFD calculations [17, 18]. However, systematic analysis of the synthesis process in RF ICP system with in-situ investigation of plasma to understand the dissociation process of precursor has not been discussed in detail. In this study, we have done extensive in situ investigation of RF inductively coupled $Ar/H_2/CH_4$ plasma by OES in order to monitor the radicals formed due to dissociation of CH_4 during the synthesis process of the graphene nano-flakes. Emission spectra of $Ar/H_2/CH_4$ plasma are dominated by the emission from C_2 Swan system. Optical emissions from $\mathrm{CH}(B^2\Sigma^- \rightarrow X^2\Pi)$, C_3, $\mathrm{CH}(A^2\Delta \rightarrow X^2\Pi)$ and H_2 are also observed. The variation of plasma and particle temperatures with different growth conditions are compared with that of production rate. The properties and quality of the synthesized graphene nano-flakes are cognized by electron microscopy, x-ray diffraction and Raman spectroscopy. The synthesized nano-flakes have uniform morphology having average length and width about 200 and 100 nm, respectively. The thickness of the nano-flakes is found to be between 3.7 to 7.5 nm, and the number of graphene layers varies from 11 to 22 depending on the growth conditions. The intensity ratio (I_D/I_G) of D and G band observed in the Raman spectra is less than 0.33 which betokens good quality of the synthesized graphene nano-flakes.

2. Experimental details

Figure 1 shows a schematic diagram of the RF inductively coupled thermal plasma facility developed at EMPA in Thun, Switzerland for the synthesis of nanoparticles. It comprises of an induction plasma torch coupled to a RF power supply, a vacuum synthesis chamber, a filtration unit, a gas manifold, a flow control system and a precursor injector. The plasma torch (PL-35, TEKNA Plasma System Inc. Canada) mounted on the top of the synthesis chamber is connected to a RF power supply (Elgotec AG, Switzerland) delivering a maximum electrical power of 35 kW at a frequency of 13.56 MHz. The plate power was varied from 12 to 18 kW in this experiment. The plasma torch consists of a four turn induction coil, a water cooled 35 mm inner diameter ceramic tube, and another 30 mm outer diameter ceramic tube which separates the central gas from the sheath gas. The central gas was introduced tangentially through a metallic gas distributor with a swirl, whereas sheath gas was injected axially along the ceramic wall. More details about the RF ICP apparatus can be found elsewhere [7–9]. Argon (Ar) gas with fixed flow rate of 12 slpm was used as central gas, and a mixture of Ar and hydrogen (H_2) gas with fixed flow rates of 60 and 4 slpm, respectively, as sheath gas. Gas precursor with carrier gas referred to as reactive gas was introduced axially into the induction plasma torch. Ar gas with fixed flow rate of 3 slpm was used as carrier gas, and methane (CH_4) gas with varying flow rates (0.3–2 slpm) was used as precursor. Thus, the total flow rates of Ar and H_2 gas were fixed at 75 and 4 slpm, respectively, during the experiment.

The water cooled synthesis chamber was connected via a filtration unit to a water-ring pump with a pressure regulation system. The pressure was varied from 400 to 700 mbar in the experiment. The synthesis chamber was also equipped with view ports for the in-situ process visualization and characterization. Optical emission spectroscopy (OES) experiment was conducted at different growth conditions using a fiber coupled spectrometer (Ocean Optics) and collecting optics through a view port located at distance of 115 mm from the exhaust of the ICP torch. Each emission spectrum was recorded in time integrated manner and spatially averaged over 5 cm^2 area of the plasma.

The filtration unit comprises of a small bypass membrane filter which was used for on-line sampling of the synthesized nanoparticles. Imaging of the synthesized nanoparticles were carried out by scanning electron microscopy (SEM) and scanning transmission electron microscopy (STEM) using a field emission microscope (Hitachi S4800 high-resolution SEM). The synthesized powders were pelletized (Ø.10 mm, thickness 1mm) for x-ray diffraction (XRD), and Raman spectroscopy characterization. XRD patterns were measured using a X-Ray Bruker D8-Discover diffractometer in Bragg–Brentano geometry. Confocal optical microscope system (NTEGRA Spectra) was used for Raman experiment. A 532 nm Nd: YAG laser with power of 10 mW was focused using a microscope objective onto the sample with a focal spot diameter of about 1 μm. The Raman scattered signal was recorded using a charge-coupled device (CCD) attached to a spectrometer in the back-scattered geometry.

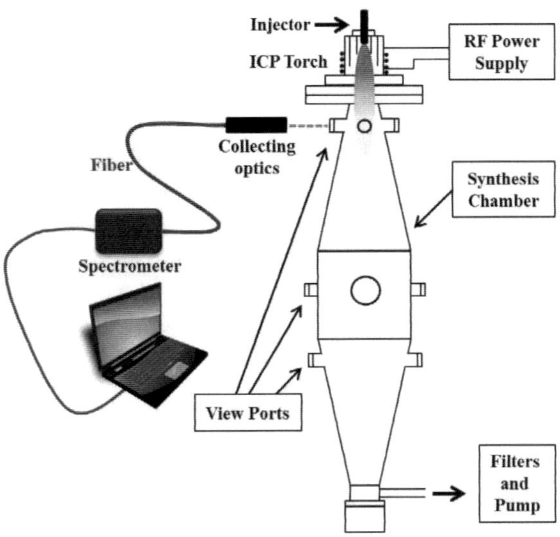

Figure 1. Schematic diagram of RF inductively coupled thermal plasma apparatus developed at EMPA in Thun, Switzerland.

3. Results and discussion

Figure 2(a) shows the optical emission spectrum of $Ar/H_2/CH_4$ plasma recorded at plate power of 18 kW, pressure of 400 mbar and CH_4 flow rate of 0.7 slpm. The observed emission bands are represented by **a-i** in the figure which corresponds to the wavelengths at 387, 405, 431, 436, 468, 516,

563, 612 and 619 nm, respectively. The C_2 Swan system corresponding to the vibration sequences $\Delta\upsilon$ = +2, +1, 0, -1, -2 of the $(d^3\Pi_g - a^3\Pi_u)$ electronic transitions dominates the optical emission spectrum which are marked by **d**, **e**, **f**, **g** and **i**, respectively. The emissions **a**, **b**, **c** and **h** at 387, 405, 431 and 612 nm, respectively, are attributed to the emissions from $CH(B^2\Sigma^- \rightarrow X^2\Pi)$, C_3, CH $(A^2\Delta \rightarrow X^2\Pi)$ and H_2 [13, 19]. No Ar I transitions are evident and atomic hydrogen transitions including emission at 656 nm (H_α) is negligibly small in comparison to the C_2 Swan band emissions. Different radicals in the emission spectra are appeared due to dissociation of CH_4. Electron impact and the dehydrogenation dissociations are the two primary mechanisms for the dissociation of CH_4 [20, 21].

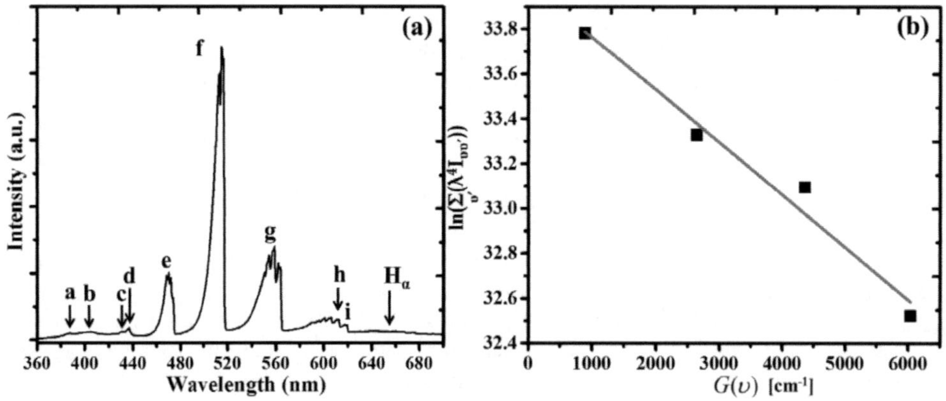

Figure 2. (a) Optical emission spectrum of $Ar/H_2/CH_4$ plasma in the spectral region between 360 and 700 nm at plate power of 18 kW, pressure of 400 mbar and CH_4 flow rate of 0.7 slpm. (b) Schematic plot between $G(\upsilon)$ and $\ln\left(\sum_{\upsilon'}\left(\lambda^4 I_{\upsilon\upsilon'}\right)\right)$ for the estimation of molecular vibrational temperature in $Ar/H_2/CH_4$ plasma at plate power of 18 kW, pressure of 400 mbar, and CH_4 flow rate of 0.7 slpm.

The observed dominant emission bands in $Ar/H_2/CH_4$ plasma originating from C_2 Swan system can be analyzed to estimate the molecular vibrational temperature using the following relation [22, 23] which is based on local thermodynamic equilibrium (LTE) condition:

$$\ln\left(\sum_{\upsilon'}\left(\lambda^4 I_{\upsilon\upsilon'}\right)\right) = C_1 - G(\upsilon)\frac{hc}{kT_{vib}} \qquad (1)$$

where λ is the wavelength corresponding to the emission, $I_{\upsilon\upsilon'}$ is the Swan band emission intensity of the progression ($\upsilon\ \upsilon'$), h the Planck's constant, c refers to the velocity of light, C_1 a constant, $G(\upsilon)$ the term value corresponding to the vibrational level in the upper electronic state and T_{vib} the vibrational temperature. The slope of the linear plot between $G(\upsilon)$ and $\ln\left(\sum_{\upsilon'}\left(\lambda^4 I_{\upsilon\upsilon'}\right)\right)$ can provide the vibrational temperature which is a measure of plasma temperature under LTE. Figure 2(b) shows a schematic plot between $G(\upsilon)$ and $\ln\left(\sum_{\upsilon'}\left(\lambda^4 I_{\upsilon\upsilon'}\right)\right)$ represented by the solid squares (■), and the solid line is the linear fit which yields a temperature of 6200 K for the $Ar/H_2/CH_4$ plasma at plate power of 18 kW, pressure of 400 mbar, and CH_4 flow rate of 0.7 slpm.

At high plate powers above 15 kW, the C_2 Swan band emissions recorded at fixed CH_4 flow rate of 0.7 slpm dominate the optical emission spectra of $Ar/H_2/CH_4$ plasma in the pressure range from 400 to 700 mbar. However, the intensity of the C_2 Swan band decreases with increase in pressure leading to the decrease in plasma temperature. The estimated plasma temperature decreases from 4060 to 3520 K at 15 kW as the pressure increases from 400 to 700 mbar. The decrease in plasma temperature with increase in pressure and decrease in plate power gives rise to the increase in condensation process of nanoparticles formation. Thus, the broad band continuum emission appears along with the C_2 Swan band in the optical emission spectra of $Ar/H_2/CH_4$ plasma at 15 kW above 600 mbar which is attributed to the emission from nanoparticles formed in vapor phase. The production rate is found to be increased from 0 to 0.3 g/h at 18 kW and from 0 to 1 g/h at 15 kW as the pressure increases from 400 to 700 mbar.

Since the plasma temperature decreases and consequently, the production rate increases for a fixed pressure and CH_4 flow rate as the plate power decreases from 18 to 15 kW, the plate power is further reduced to 12 kW in order to see the effect of decreasing plasma temperature on the optical emission characteristics and the production rate at different pressures. The optical emission spectra of $Ar/H_2/CH_4$ plasma at plate power of 12 kW and CH_4 flow rate of 0.7 slpm at all pressures are found to be dominated by the continuum emission from the nanoparticles, and the C_2 Swan band emission is negligible as shown in Fig. 3. The particle temperature is estimated from the spectral response corrected continuum emission following Ref. **24** which decreases from 2200 to 1700 K with increase in pressure from 400 to 700 mbar similar to the behavior of plasma temperature. The production rate is also increased with increase in pressure which is consistent with that of at 15 and 18 kW. In addition, the production rate is more at 12 kW than that at 15 and 18 kW for a fixed pressure and CH_4 flow rate due to increasing condensation of supersaturated vapor at lower temperature. The combination of lowest plate power (12 kW) and highest pressure (700 mbar) considered in this experiment yields higher production rate of 1.7 g/h than other combinations of plate powers and pressures at fixed CH_4 flow rate of 0.7 slpm. Thus, we varied the CH_4 flow rate at fixed pressure of 700 mbar and plate power of 12 kW in order to optimize the CH_4 flow rate to achieve maximum production rate. However, we could not conduct the experiment at 12 kW and 700 mbar at higher CH_4 flow rates than 0.7 slpm since the capacity to dissociate CH_4 and produce plasma was insufficient. Therefore, the plate powers were increased to 15 and 18 kW at 700 mbar and the production rate was monitored at different CH_4 flow rates. The production rate is found to be maximum at 1.5 slpm of CH_4 flow rate for both plate powers, and is 6 and 7 g/h at 15 and 18 kW, respectively.

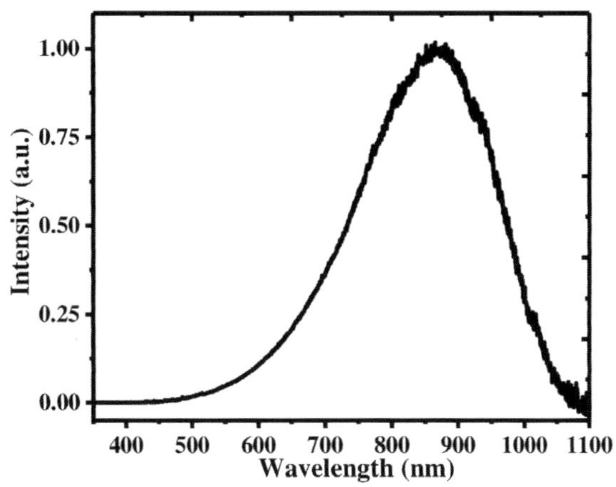

Figure 3. Typical broad band continuum emission from the nanoparticles formed in vapor phase in $Ar/H_2/CH_4$ plasma at plate power of 12 kW, pressure of 700 mbar and CH_4 flow rate of 0.7 slpm.

The nanoparticles collected at the filter were characterized by SEM and STEM in order to understand the surface morphology. Figure 4 shows the representative SEM and STEM images of the nanoparticles which depicts uniform surface morphology. The nanoparticles exhibit flake like structures having average length and width of about 200 and 100 nm, respectively, irrespective of the growth conditions. Figure 5(a) shows the XRD spectrum of the nano-flakes synthesized at plate power of 18 kW, pressure of 700 mbar, and CH_4 flow rate of 0.7 slpm which is a representative spectrum for all growth conditions. The diffraction pattern shows the crystalline nature of the nano-flakes. Four diffraction peaks are identified at 2θ angles of 26°, 43°, 54° and 78°. The diffraction peaks at $2\theta = 26°$ and 43° correspond to the (002) and (100) planes, respectively, which are the typical features of the graphitic structure and have been identified in XRD spectrum of graphene sheets. The peak at $2\theta = 54°$ corresponds to the (004) plane parallel to the (002) plane at $2\theta = 26°$ of the graphene layers. In addition, the diffraction lines at $2\theta = 43°$ and 78° corresponding to the planes (100) and (110), respectively, are the characteristics of the 2D in-plane symmetry along the graphene layers [25]. The thickness of the graphene nano-flakes is estimated using the Scherrer formula, and the layer-to-layer distance (d_{002}) between two successive graphene layers is calculated using the peak position of the (002) plane [26]. The number of layers in each graphene nano-flakes sample is determined by using the values of corresponding thickness and d_{002} [25]. The thickness of the nano-flakes is found to be between 3.7 to 7.5 nm and the number of layers in the synthesized graphene nano-flakes varies between 11 and 22 depending on the growth conditions.

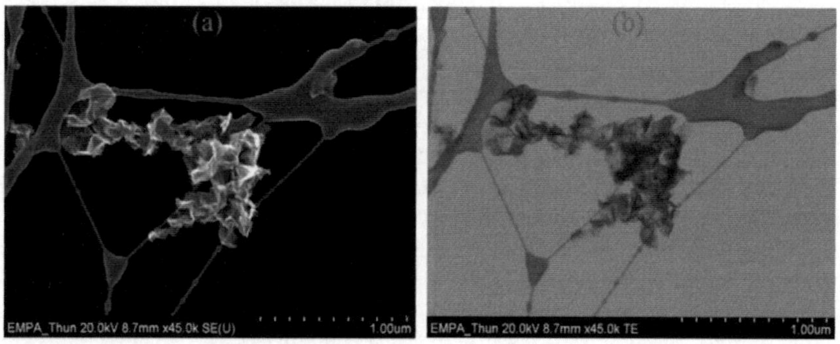

Figure 4. SEM (a) and STEM (b) images of the synthesized nano-flakes at plate power of 15 kW, pressure of 700 mbar and CH_4 flow rate of 0.7 slpm.

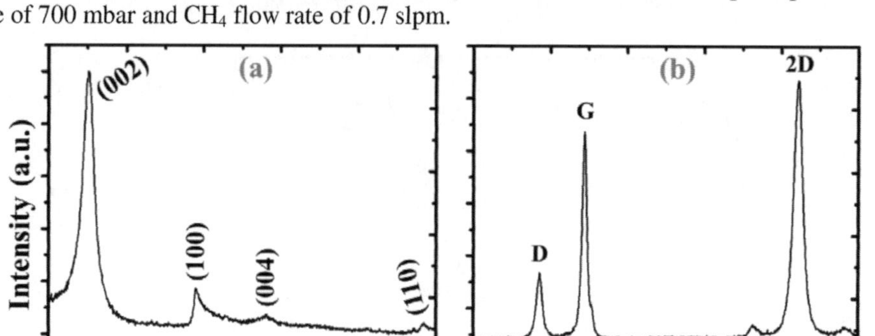

Figure 5. Representative XRD (a) and Raman (b) spectra of nano-flakes synthesized at plate power of 18 kW, pressure of 700 mbar and CH_4 flow rate of 0.7 slpm.

Raman spectroscopy was used to evaluate the quality of the nano-flakes synthesized at different growth conditions. Figure 5 (b) shows a representative Raman spectrum of the synthesized nano-flakes at plate power of 18 kW, pressure of 700 mbar and CH_4 flow rate of 0.7 slpm. The main features observed in the Raman spectra of all samples are D band at 1345 cm^{-1}, G band at 1577 $cm^{-1}$and 2D band at 2686 cm^{-1}. The D band at 1345 cm^{-1} does not appear in pristine graphene and is observed where symmetry is broken by edges or in samples with a high density of defects [16, 27]. Since the average length and width of investigated graphene nano-flakes are about 200 and100 nm, respectively, major contribution of edge effects to the D band intensity is expected. The intensity ratio of D and G band (I_D/I_G), a measure of the quality of the material, is found to be less than 0.33 in the synthesized graphene nano-flakes for all growth conditions. The smaller value of I_D/I_G ratio (< 0.33) in spite of having contribution from edges to the D band intensity ensures high quality nano-flakes in our case. The variation of I_D/I_G ratio with different growth conditions is also found to be negligibly small i.e. from 0.24 to 0.33 which indicates no influence of processing conditions on the quality of the synthesized graphene nano-flakes. Moreover, the synthesized graphene nano-flakes have smaller values of I_D/I_G ratio in comparison to microwave plasma enhanced chemical vapor deposited graphene nano-flakes of similar width [28] which further ensures quality growth at low cost with higher production rate in the ICP synthesis system.

4. Conclusions

Graphene nano-flakes are synthesized by using RF ICP synthesis system at different growth conditions with in situ investigation of $Ar/H_2/CH_4$ plasma by optical emission spectroscopy. The C_2 Swan system corresponding to the vibration sequences $\Delta\upsilon$ = +2, +1, 0, -1, -2 of the $(d^3\Pi_g - a^3\Pi_u)$ electronic transitions dominates the optical emission spectra. The molecular vibrational temperature, a measure of plasma temperature, estimated by analyzing the C_2 Swan band emission intensities is found to be decreased with increase in pressure and decrease in plate power. The production rate is also increased with increase in pressure and decrease in plate power. Thus, the increase in production rate is associated with decrease in plasma temperature due to increasing condensation of supersaturated vapor. The broadband continuum emission appears in the optical emission spectra at certain growth conditions in which formation of vapor phase nanoparticles is facilitated. Thus, continuum emission acts as an indicator of the vapor phase nanoparticles formation. The maximum production rate of the graphene nano-flakes synthesized in current ICP apparatus is 7 g/h at plate power of 18 kW, pressure of 700 mbar and CH_4 flow rate of 1.5 slpm. The capacity of RF ICP synthesis system can further be improved to increase the production rate at lower processing cost. Higher plate power and pressure together with higher CH_4 flow rate could probably lead to higher production rate in the ICP synthesis system. The synthesized graphene nano-flakes have uniform morphology having average length and width about 200 and 100 nm, respectively, irrespective of the growth conditions. Raman analysis reveals that the synthesized graphene nano-flakes are of high quality since the I_D/I_G ratio, a measure of quality of the material, is found to be less than 0.33 in spite of having contribution from edges of the nano-flakes to the D band emission intensity (I_D). Since the variation of I_D/I_G ratio with growth conditions is insignificant, the observed variation of plasma or particle temperature with growth conditions is uncorrelated with the quality of the synthesized graphene nano-flakes.

Acknowledgements

The authors would like to thank European Union's H2020 research and innovation programme for two EU projects: SPICY (grant no. 653373) and NANOLEAP (grant no. 646397). Authors would like to thank James Best for his help in Raman experiment.

References

[1] Byrappa K, Ohara S and Adschiri T 2008 *Adv. Drug Delivery Rev.* **60** 299
[2] Mohanta A and Thareja R K 2009 *J. Appl. Phys.* **106** 124909
[3] Mohanta A, Kung Patrick and Thareja R K 2015 *Appl. Phys. Lett.* **106** 013108
[4] Patel D N, Pandey P K and Thareja R K 2013 *Appl. Opt.* **52** 7592
[5] Jiayin G, Xiaobao F, Dolbec R, Siwen X, Jurewicz J and Boulos M 2010 *Plasma Sci. Tech.* **12** 188
[6] Boulos M I 1997 *High Temperature Material Processes* **1** 17
[7] Shin J W, Miyazoe H, Leparoux M, Siegmann St, Dorier J L and Hollenstein Ch 2006 *Plasma Sources Sci. Technol.* **15** 441
[8] Leparoux M, Schreuder C, Shin J-W and Siegmann S 2005 *Adv. Eng. Mater.***7** 349
[9] Leparoux M, Schreuder C and Fauchais P 2008 *Adv. Eng. Mater.* **10** 1147
[10] Pristavita R, Meunier J-L and Berk D 2011 *Plasma Chem. Plasma Process* **31** 393
[11] Bashir M, Rees J M, Bashir S and Zimmerman W B 2014 *Phys. Lett. A* **378** 2395
[12] Ma J, Ashfold M N R and Mankelevich Y A 2009 *J. Appl. Phys.* **105** 043302
[13] Zhou H, Watanabe J, Miyake M, Ogino A, Nagatsu M and Zhan R 2007 *Diamond Relat. Mater.* **16** 675
[14] Mohanta A, Singh V and Thareja R K 2009 *J. Appl. Phys.* **104** 064903
[15] Kushwaha A, Mohanta A and Thareja R K 2009 *J. Appl. Phys.* **105** 044902
[16] Ferrari A C and Basko D M 2013 *Nature Nanotechnol.* **8** 235
[17] Pristavita R, Mendoza-Gonzalez N-Y, Meunier J-L and Berk D 2011 *Plasma Chem Plasma Process* **31** 851
[18] Meunier J-L, Mendoza-Gonzalez N-Y, Pristavita R, Binny D and Berk D 2014 *Plasma Chem Plasma Process* **34** 505
[19] Denysenko I B, Xu S, Long J D, Rutkevych P P, Azarenkov N A and Ostrikov K 2004 J. Appl. Phys. **95** 2713
[20] Luque J, Juchmann W, Brinkman E A and Jeffriesc J B 1998 *J. Vac. Sci. Technol. A* **16** 397
[21] Heintze M, Magureanu M and Kettlitz M 2002 *J. Appl. Phys.* **92** 7022
[22] Harilal S S, Issac R C, Bindhu C V, Nampoori V P N and Vallabhan C P G 1997 *J. Phys. D: Appl. Phys.* **30** 1703
[23] Chehrghani A and Torkamany M J 2013 *Opt. Laser Eng.* **51** 61
[24] Amoruso S, Ausanio G, Bruzzese R, Vitiello M and Wang X 2005 Phys. Rev. B **71** 033406.
[25] Dervishi E, Li Z, Watanabe F, Biswas A, Xu Y, Biris A R, Saini V and Biris A S 2009 *Chem. Commun.* **27** 4061
[26] B. D. Cullity and S. R. Stock, Elements of X-Ray Diffraction, 3rd ed. Prentice-Hall, Englewood Cliffs, NJ, 2001.
[27] Allen M J, Tung V C and Kaner R B 2010 Chem. Rev. **110** 132
[28] Deng J-H, Deng L-N, Liu R-N, Han A-L, Li D-J and Cheng G-A 2016 *Carbon* **102** 1

IOP Conf. Series: Journal of Physics: Conf. Series **825** (2017) 012011 doi:10.1088/1742-6596/825/1/012011

Plasma-arc production of synthetic rutile from leucoxene concentrate

A.A.Nikolaev, D.E.Kirpichev, A.V.Nikolaev, A.V.Samokhin

Federal state budgetary institution of science Institute of metallurgy and materials science of A.A. Baykov of the Russian Academy of Sciences (IMET RAS), Moscow, Leninskiy av., 49, Russian Federation

dym2004@bk.ru, nikolaev@imet.ac.ru

Abstract. Plasma production of synthetic rutile from leucoxene concentrate with high contents of SiO_2 is described. Concentrate was melted with carbon in laboratory plasma-arc furnace. Synthetic rutile with 3.6 % Si was produced. Experimental results and thermodynamic calculations are presented.

1. Introduction

The main Russian titanic ore resources exist in the form of the leucoxene oil-bearing sandstone which consists mainly of SiO_2 and contains TiO_2 up to 10 % (Yarega deposit). The enrichment of leucoxene ore is carried out in few stages. It is distillation from oil first of all and then ore is subjected to flotation. The concentration of TiO_2 raises up to 40-50 %. Further enrichment to 80-85% of TiO_2 is performed in autoclaves by treatment with hot H_2SO_4. The autoclave treatment is complicated and ecologically dangerous [1-3]. The aim of the article is to show that the autoclave treatment can be replaced by carbon reduction without dangerous reagents. The essence of the process is reduction of SiO_2 to SiO. SiO will evaporate from concentrate at high temperature. The high temperature can be obtained by plasma-arc heating [4].

2. Thermodynamic calculation

The model system for thermodynamic calculation contained 0.75 mole TiO_2, 1 mole SiO_2 and C. The amount of C was varied from 1 up to 4 moles. Calculation was executed by means of the program complex TERRA [5, 6] at pressure p = 0.1 MPa and temperature T = 1000 – 4000 K. It was considered that the condensed phase exists as oxide solution without heat of solution. The results are shown in figures 1 and 2. Figure 1 illustrates the composition of the system when amount of C equals 1 mole. The evaporation of SiO begins when temperature is over 2000 K. Almost all SiO_2 in system is reduced and evaporated when temperature is 3000 K. It is not good to heat the system more because the evaporation of titanium oxides also begins.

Not all amount of carbon in system is spent for reduction of SiO_2. Some part of carbon is spent for reduction of TiO_2 to Ti_xO_y. It is useful to have 1.5 moles of carbon in the system for this reason. More carbon is not good because of formation of Si, SiC, TiSi which are not volatile. Figure 2 illustrates dependence of ε_{Si} upon contents of carbon and temperature of the system. ε_{Si} is the ratio of amount of Si in gas phase to total amount of Si in system. According to Figure 2 the optimal conditions are: C = 1.5 mole, T = 2600 K. Synthetic rutile contains 3 mass % of Si in this case. Gas phase contains ≈ 44 mass % CO and ≈ 56 mass % SiO, $\varepsilon_{Ti} < 0.01$. Theoretical energy consumptions (without losses) are 8.9 GJ/t of the concentrate.

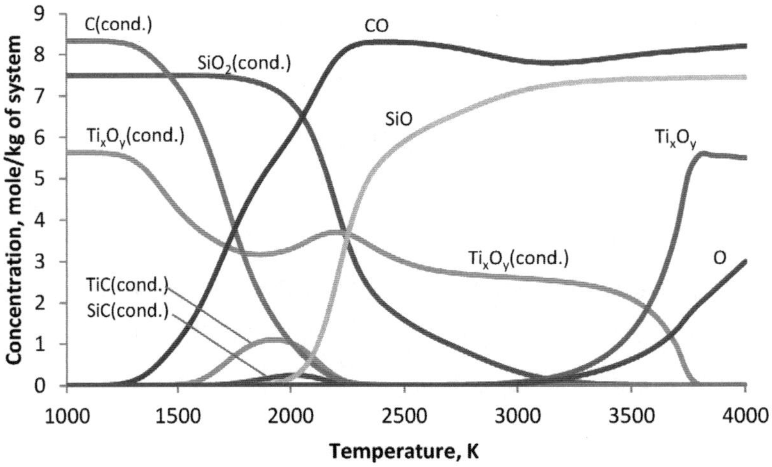

Figure 1. Equilibrium mixture of system 0.75 TiO_2 + SiO_2 + C calculation results

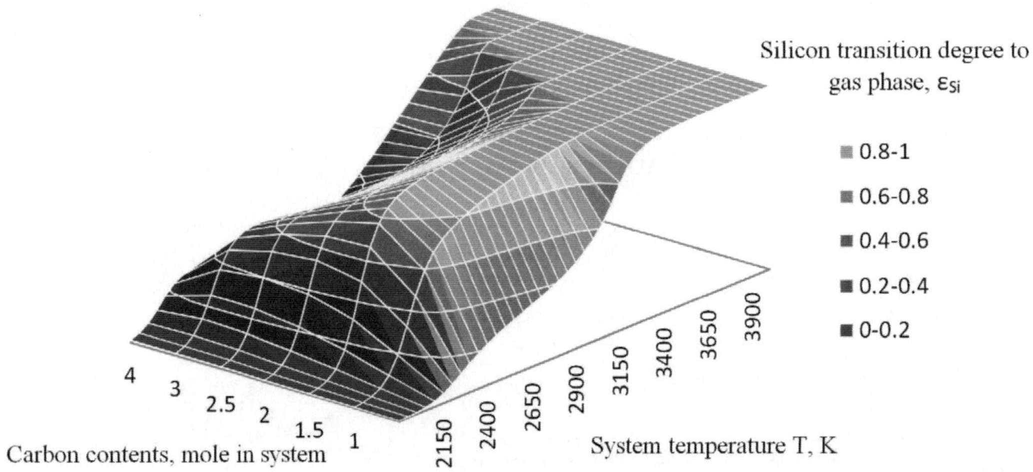

Figure 2. Silicon transition degree to gas phase ε_{Si} in dependence of system temperature T and carbon contents

3. Experiment technique

Experiments were executed at atmospheric pressure in laboratory DC plasma-arc furnace (figure 3). The furnace included copper cylindrical water cooled camera (1) with internal diameter 100 mm. Camera was closed from above by lid (2) through which graphite electrode (4) was entered by means of copper holder (3). Diameter of electrode was 50 mm. The lid was separated from camera by isolators (5) and gas collector (6). The gas collector served for supplying of argon under lid. The electrode was connected to a negative pole of electric source, the camera to a positive. The bottom of the camera was protected from extreme heating by graphite disc (7).

The electrode (4) had possibility of axial movement. Plasma arc (8) started when electrode (4) and disc (7) are brought in contact. The current of the arc was regulated by ballast resistance (9). Electrode

(4) and holder (3) had axial channels through which concentrate-carbon mixture was fed by feeder (10). The concentrate mixture formed the molten pool (11) while melting under arc heating.

There was an electromagnetic winding on the external surface of camera (1) (it is not shown on the figure), by means of which axial magnetic field ≈ 100 kA/m was created. The field promoted spatial stabilization of arc.

The sidewall of the camera (1) was equipped with window (12) for supervision over the process. The window was flushed by argon for prevention of a dust accumulation.

Argon was fed into electrode together with concentrate mixture. Argon served as transporting and plasma-forming gas. Hot gas products of the process (SiO, CO) mixed with Ar were transported to the heat exchanger (13) in which they were exposed to cooling. The cooled gas came to filter (14). Filter (14) was made as a combination of glass fabric and a steel grid of volume weaving. The total consumption of argon was 0.6 g/s: 70% in electrode, 15% on window, 15 % under lid.

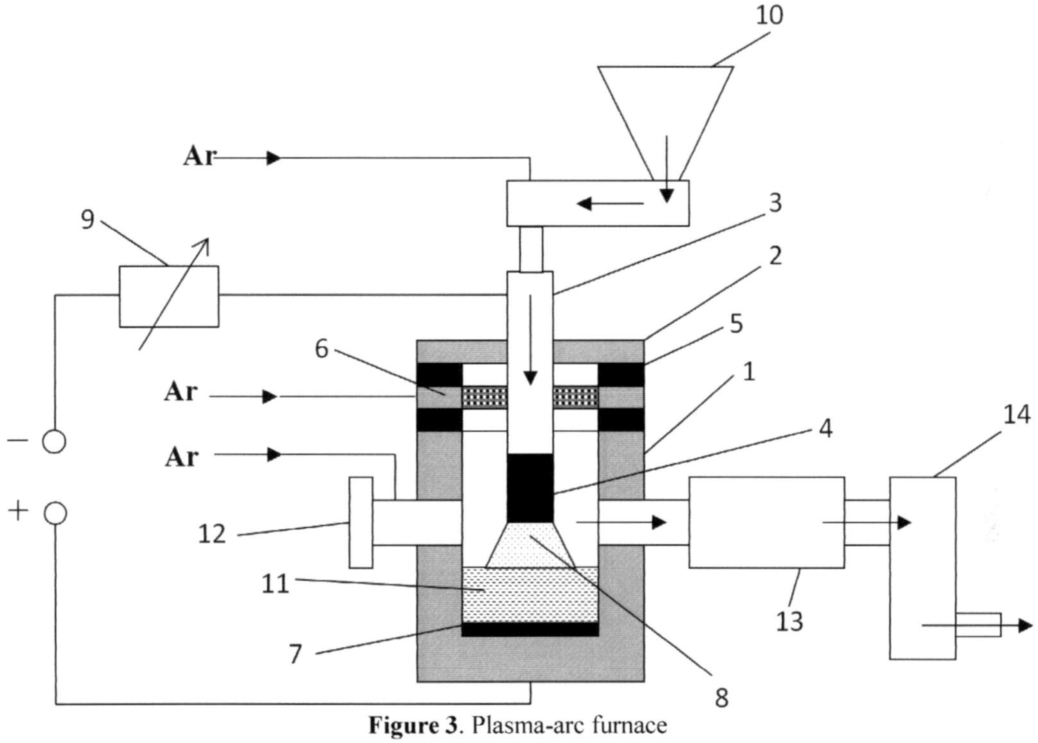

Figure 3. Plasma-arc furnace

Two condensed products were received after each experiment: ingot and powder. Ingot located on the bottom of camera, powder located on the upper part of camera, in the filter and in the heat exchanger. The condensed products were collected and weighed. It was not noticed that graphite disc (7) reacted with melt (11). The difference of mass of charged concentrate mixture and condensed products was thought to be a gas. It was assumed that this gas consists mainly of CO. Only powder from filter was analyzed. Its amount was about 10 % from total mass of powder. It was considered that this powder was not contaminated with initial concentrate.

The analysis of the Ti and Si concentration in products was carried out by atomic emission method with the inductive coupled plasma (AES-ICP) on spectrometer ICAP-6500 Duo (Thermo Scientific, USA) with the software of ITEVA. For definition of elements a standard sample Merck ICP multi element standard solution XVI was used.

The composition of the condensed powder was determined by measurement of a specific surface by isotherms of adsorption of gas on a surface of samples. For definition of nanopowders specific surface the analyzer of a specific surface TriStar 3000 was used (Micromeritics, the USA).

The morphology of ingot and powder material was analyzed on optical and electronic microscopes: on raster dual-beam microscopes of Helios 650 and Versa 3D (FEI Company), and also on the translucent electronic microscope of Tecnai G2 F20 (FEI Company) with adapter for the power dispersive x-ray analysis of a chemical composition.

The phase composition was examined by method of the X-ray phase analysis on the Ultima-4 diffractometer (Rigaku, Japan) with use of the program complex MAUD developed on the basis of the modified Ritveld's method and PDXL with the PDF-2 database (Powder Difraction File).

The carbon contents in samples was measured on the CS-600 analyzer (Leco, USA) by method of burning of samples in oxygen flow with the subsequent definition of the carbon contents in gas phase by means of infrared spectrometry.

Camera (1), lid (2), holder (3) and heat exchanger (13) (figure 3) were cooled with water and calorimetred. The temperature of gas after heat exchanger was near the room temperature

4. Experiment results and its discussion

Concentrate was a loose oxide material with a size of particles ≈ 1 mm. Chemical composition: Si-17.0, Ti-32.8, Al-1.65, Fe-2.07 mass %. Chemically pure carbon (%C > 99.9%) with size of particles 0.3 - 0.5 mm was used as a reducing agent. Charge mixture was prepared on Turbula C20 setup. Three experiments differing in amount of carbon in charge mixture and plasma arc power were carried out. The parameters of experiments and properties of products are presented in table 1.

Table 1. Parameters of experiments and properties of products

Parameters	Melting No.1	Melting No.2	Melting No.3
Contents of carbon in charge mixture, mole C/mole Si	1.8	3.4	3.4
Plasma arc power, kW	10.3	10.0	18.5
Plasma arc current, A	332	334	691
Charged mass of concentrate mixture, g	437	327	327
Masses of products, g			
- ingot	202	148	150
- powder	103	72	73
- gas (calculated as difference)	132	107	104
Content in ingot, mass.%			
Si	8.8	7.3	3.6
C	0.9	10.0	9.12
Content in powder from filter, mass.%			
Si	33.7	45.7	47.5
C	3.3-7.24	6.29	4.05

The content of silicon in ingots was significantly less than in an initial concentrate in all experiments. The increase of contents of carbon in charge mixture in melting No.2 and power of plasma arc in melting No.3 promoted desiliconization of concentrate. So, in melting No.3 the contents of silicon in the ingot was 3.6% against 8.8% in melting No.1. The contents of silicon in powder increased on the contrary.

Analysis of ingot on optical microscope has shown the existence of two phases. The microstructure of ingot is given in figure 4. Dark fields correspond to silicon oxide, light – to titanium oxide. The eutectic on basis of titanium oxide is distinguishable: areas containing dot inclusions of SiO_2.

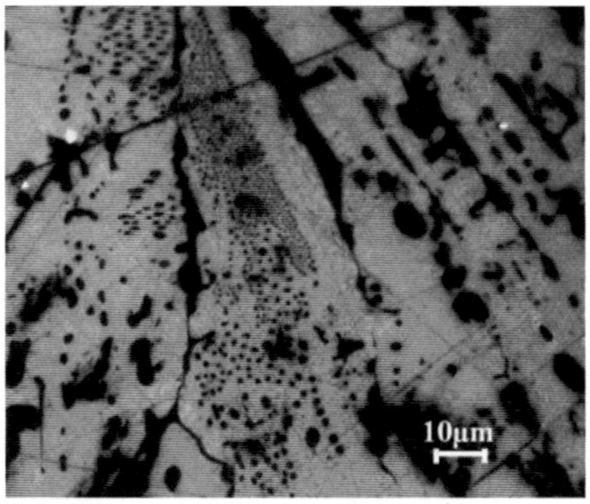

Figure 4. Microstructure of ingot

There are lines of quartz and rutile on the X-ray pattern of initial concentrate, whereas peaks of TiO_2 on the X-ray pattern of the ingot from melting No.1 are not found (figure 5). But there are peaks of less oxidized compounds: Fe (TiO_3) and Fe (Ti_2O_5). Quartz peaks in this sample are absent also. It can be explained by SiO_2 amorphization when hardening.

There are no peaks of Fe-bearing compounds on X-ray pattern of initial concentrate because these compounds have no peaks in realized range of θ.

There are no peaks of Fe (TiO_3) and Fe (Ti_2O_5) on X-ray patterns of the ingots from melting No.2 and No.3 at high contents of carbon in charge mixture. However there are TiO peaks which are absent on X-ray patterns of initial concentrate and ingot from melting No.1. Thus, the increase of carbon contents in charge mixture leads to titanium oxides reduction.

The peaks of Ti-bearing compounds are not revealed on X-ray patterns of powder. The pattern of melting No.1 with moderate amount of carbon contains the Si and SiO_2 peaks. SiO_2 peak disappears in pattern for melting No.2 with more amount of carbon in charge mixture. SiC peak appears in pattern for melting No.3 with high amount of carbon and high power of plasma arc.

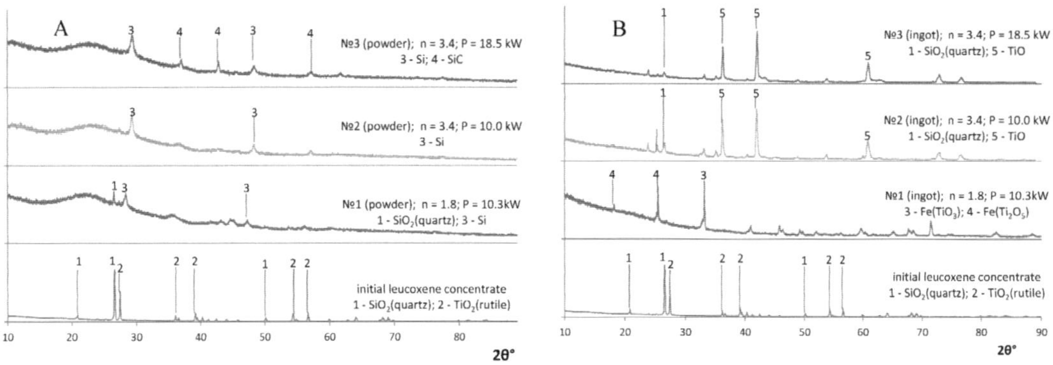

Figure 5. X-ray patterns of initial concentrate and melt products. A-powder; B-ingot; n – contents of C in concentrate mixture, mole C/mole Si

The powder had microstructure as if it had condensed from gas phase. It was presented generally by fibres about 10 nanometres thick and up to 1 micron long (figure 6). The specific surface of powder was ≈ 70 m^2/g. There were spherical particles in size of 20 – 100 nanometres (figure 7). Fibres and spherical particles had layered structure with impurity of Al and Fe. Spherical particles, for example, could have the two-layer structure including the kernel enriched with Ti. The upper layer of the particle is enriched with Si (figure 8). The contents of silicon in some particles is significantly higher than deduced from the results of chemical analysis. The reason is: chemical analysis gives the average structure of great amount of particles. So, separate groups of particles might contain up to 90 mass.% Si. Thus, the contents of Si in various particles can significantly differ. The contents of impurities (Fe, Al) is about several percents at figure 8. It coordinates with the results of chemical analysis: 7.11 % Fe and 1.6 % Al for melting No1.

Figure 6. The typical microphotography of condensed powder

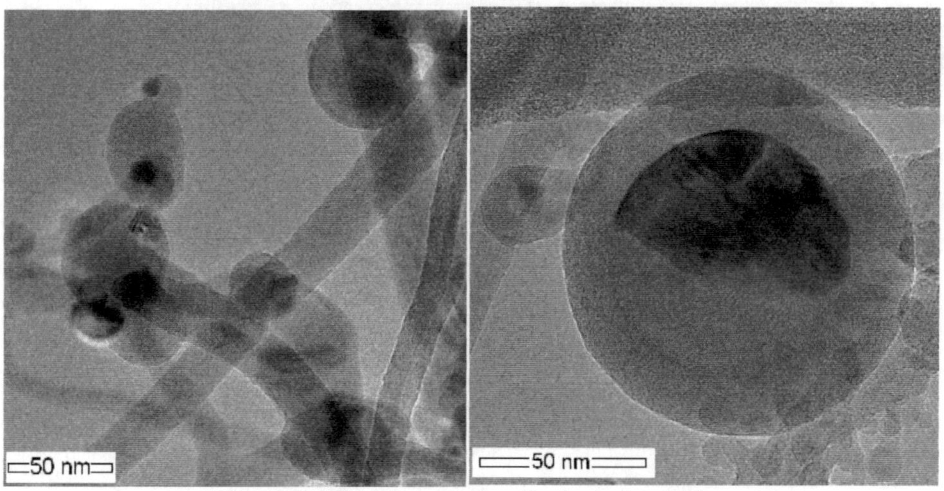

Figure 7. Microphotographs of spherical particles

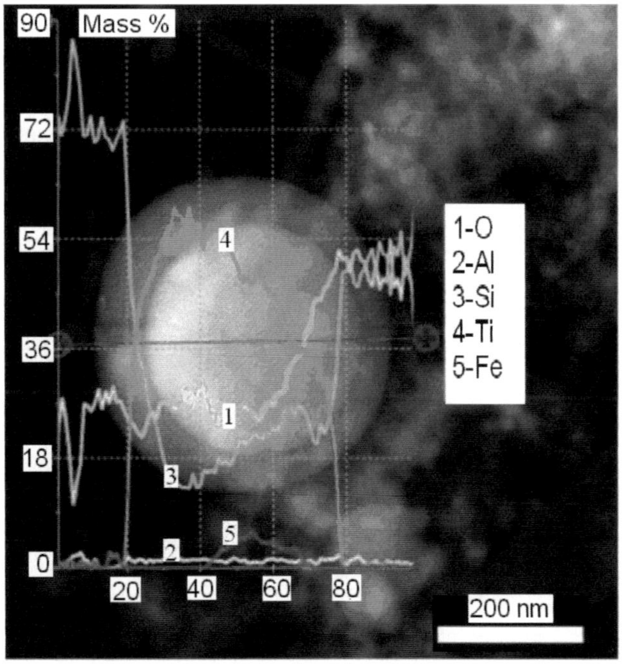

Figure 8. Composition of a spherical particle

The power of plasma arc exceeded the power of cooling at size ΔQ ≈ 2,5 kW. This testifies that endothermic processes took place in the furnace: formation of oxide melting pool ≈ 1,5 kW and formation of siliceous fraction ≈ 1 kW.

5. Conclusions

The possibility of plasma-arc production of synthetic rutile from leucoxene concentrate is shown. Synthetic rutile with 3.6% of silicon was produced in laboratory experiments at energy consumption 20 - 40 GJ/t of concentrate and specific productivity of plasma furnace 20 t/(m3·day). The additional product in the form of ultra-disperse powder on the base of silicon was produced. The base of powder is fibers 10 - 50 nanometers thick and about 1 micron long.

The optimal thermodynamic conditions of plasma-arc reduction of leucoxene concentrate are: C = 15 % in mixture with concentrate, T = 2600 K. Synthetic rutile contains 3 % of Si in this case. Gas phase contains ≈ 44 % CO and ≈ 56 mass % SiO. Theoretical energy consumptions (without losses) are 8,9 GJ/t of concentrate.

References

[1] Sadihov G.B. Novy`e podhody` k resheniiu problem ispol`zovaniia kompleksnogo titanovogo i drugikh vidov trudnoobogatimogo sy`r`ia Rossii [New methods of solving the problems of using complex titanium and other difficult in enrichment raw materials]. In the book “The Baikov Institute of metallurgy and materials science 75 years”, Moscow, Intercontact Nauka, 2013, p. 37 - 59 (In Russ.).

[2] Reznichenko V.A., Ustinov V.S., Kariazin I.A., Petrunko A.N. E`lektrometallurgiia i himiia titana [Electrometallurgy and chemistry of titanium], Moscow, Nauka, 1982, 280 p. (In Russ.).

[3] Zanaveskin K.L., Maslennikov A.N., Mahin M.N., Zanaveskin L.N. Osobennosti himicheskogo i mineral`nogo sostava chernovogo kvartc-lei`koksenovogo koncentrata iaregskogo mestorozhdeniia [The peculiarities of chemical and mineralogical composition of rude

quarts-leucoxene concentrate of Yarega deposit]. Enrichment of ores, 2015, № 5, p. 25 – 32 (In Russ.)

[4] Nikolaev A.V., Nikolaev A.A., Kirpichev D.E., Samokhin A.V. Sposob pererabotki lei`koksenovogo koncentrata i ustroi`stvo dlia ego osushchestvleniia [The mode of treatment of leucoxene concentrate and apparatus for its realization]. Russian patent No 2586190, 2016 (In Russ.)

[5] Vatolin N.A., Moiseev G.K., Trusov B.G. Termodinamicheskoe modelirovanie v vy`sokotemperaturny`kh neorganicheskikh sistemakh [Thermodynamic modeling of high temperature nonorganic systems]. Moscow, Metallurgia, 1994, 352 p. (In Russ.)

[6] Trusov B.G. Komp`iuternoe modelirovanie fazovy`kh i himicheskikh ravnovesii` [The program system of modeling of phase and chemical equilibriums]. ISSN 0236-3933, Vestnik MGTU im. N.E. Baumana, ser. "Priborostroenie" , 2012, p. 240 -249 (In Russ.)

IOP Conf. Series: Journal of Physics: Conf. Series **825** (2017) 012012 doi:10.1088/1742-6596/825/1/012012

Experimental investigations on arc movement and commutation in the Low-Voltage Circuit Breaker

Jean Quéméneur[1,*], Mathieu Masquère[1], Pierre Freton[1], Jean-Jacques Gonzalez[1], Patrice Joyeux[2]

[1] Université de Toulouse ; UPS, CNRS, INPT ; LAPLACE (Laboratoire Plasma et Conversion d'Energie) ; 118 route de Narbonne, F-31062 Toulouse cedex 9, France
[2] Hager Electro SAS, 132 boulevard d'Europe, BP3, 67210 Obernai, France
[*] E-mail: quemeneur@laplace.univ-tlse.fr

Abstract. Displacement of an electrical arc between two parallel arc runners is studied with high-speed imaging, voltage, current and pressure measurements. The experimental set-up is detailed from the current source to the post-treatment methods. The influence of intensity, gas exhausts and volume behind the moving contact on arc commutation and movement are investigated. In the context of Low-Voltage Circuit Breaker (LVCB), it is shown that for high-intensity arcs, evacuating the hot gas is necessary for a good motion of the arc. Furthermore, a configuration of exhaust allowing a quick movement will not work for all current levels.

1. Introduction

A proper low-voltage breaking goes through the fast movement of the arc from its ignition point to the splitter plates that are responsible for arc extinction by current limitation. One very important force acting upon this displacement is the Laplace force due to the current circulation. In some devices this magnetic force is improved in order to accelerate the arc [1-3].

Another way to speed-up this movement is to design the chamber so that the pressure formation caused by the electrical arc would lead the gas to flow toward the splitters. On the study of McBride & al. [4], the volume behind the moving contact is obstructed in order to accelerate the commutation of the arc from the contact to the arc runner. A numerical study [5] points out that a smaller volume behind the point of arc ignition tends to improve the motion. Influence of the width of the chamber has been studied by Lindmayer & Paulke [6]. It is shown that the chamber must be wide enough to allow the hot gases created in front of the arc to get behind it in a double vortex movement. Otherwise, pressure ahead the arc rises and slows down its displacement. The effect of gas exhausts on arc motion and pressure formation is also investigated. As reported by Berger & al. [7], exhaust of the gas through slits along the arc movement leads to a faster arc movement. The necessity of an opening in the downstream is highlighted by Zeller & Rieder [8]. Reducing the downstream exhaust can lead to immobility of the arc.

All those studies carry valuable clues to understand the arc behaviour but sometimes lacks the statistic treatment required on such erratic phenomena or could use a wider range of measurements. To complete those works, we wanted to perform a statistic investigation on parameters such as exhausts configuration, chamber volume and current level with high-speed imaging, current, voltage and pressure measurements. The aim of this paper is to present a new experimental approach of arc movement and to investigate the impact of gas exhaust and of the upstream volume.

2. Experimental set-up

In order to reproduce the principle of the LVCB we created a device with a simplified arc chamber and a mechanism allowing the opening of the mobile contact at a specified time and speed. This system is synchronised with the pulse current source so that the contact opening is made at a specified current while measurements of current, voltage, pressure and high-speed imaging are performed.

2.1. Pulse current generator

To reproduce a short-circuit fault, a pulse current generator has been designed. A bench of capacitors is charged at 600 V and discharges through air coils thanks to power thyristors. This creates a 50 Hz half-sine with a prospective current of up to 13 kA. The intensity level is changed by using another inductor-capacitor combination or by lowering the charging voltage.

2.2. Arc chamber

In industrial Miniature Circuit Breakers (MCB), the arc chamber has adopted a complex shape for the sake of space saving. Therefore, a lot of geometric parameters appear which makes difficult the analysis of the arc behaviour. For this reason, we used a simplified geometry (Figure 1) where the arc chamber is a rectangular box with two parallel arc runners and a displacement of the arc in only one dimension. We also removed the splitter plates as we focus on arc movement and not on the extinction phase.

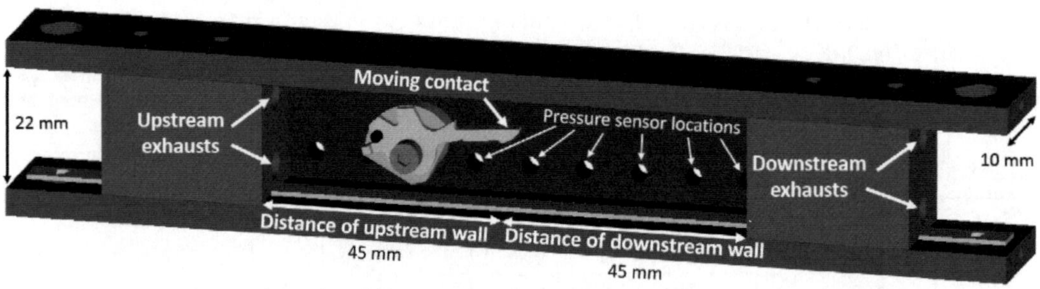

Figure 1: View of the simplified arc chamber

There are less geometric parameters with our device which are also easily modifiable, like the width and length of the chamber, the volume behind the moving contact or the number of gas exhausts of the chamber. Eight positions in the back of the chamber allow pressure measurements along the arc movement. The front wall of the chamber is made of transparent Plexiglas which permits imaging of the arc. In the middle of the chamber is an electrical contact enabling arc ignition by contact opening with a rotation like in MCBs. A dedicated mechanism has been designed in order to perform this opening movement at specified time and speed.

2.3. Opening mechanism

We need an instantaneous opening speed among 1 and 10m/s to investigate the impact of this parameter on arc breaking. Also, a good synchronisation with the current pulse is necessary to ensure arc ignition is performed with the same current. Such an actuator is not commonly found so we conceived our own system.

To achieve this rotation, a lever is hit at high velocity by a mass pulled by a spring and liberated by a lock (Figure 2). For the synchronisation, an optocoupler detects the mass movement before it hits the lever. This optical sensor activates a timer which triggers both the current pulse and the measurements.

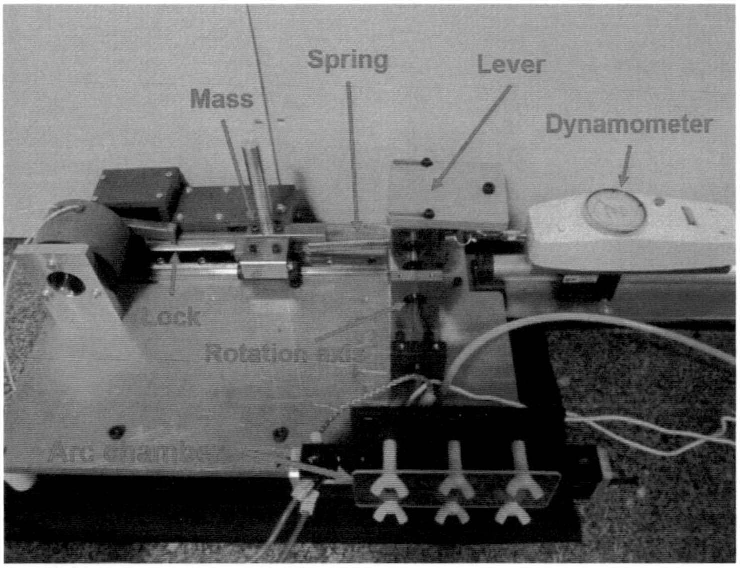

Figure 2: View of the opening mechanism

2.4. Measurements

Several electrical signals are acquired during the breaking. The arc voltage is measured at 500 kHz and reveals important information such as arc ignition or restrikes. It is also necessary to calculate the energy dissipated in the discharge. Two currents are acquired with Rogowski coils: the total current and the current flowing in the mobile contact. Knowing those two data, we can determine when the current is transferred from the moving contact to the upper arc runner in a process called commutation.

In addition with electrical measurements, pressure inside the chamber is measured at several points in the back of the chamber. We need a good time resolution and to install multiple sensors in a small volume. So, we used the Keller M5HB miniature piezoresistive sensor which has a 6.2mm diameter for the body and a dynamic response of up to 50 kHz. Moreover, it is not sensible to the shock induced by the opening mechanism. A copper muffler is used to protect the sensor head from heat and to remove the dynamic component of the pressure.

The arc movement is quantified thanks to the Photron SA5 high-speed camera with an acquisition frequency of 100,000 frame/second and a spatial resolution about 0.19 mm/pixel. To extract a relevant information of position of the arc from those images we need to develop a method.

2.5. Post-treatment

The arc and arc roots positions are sometimes not easy to determine from the pictures of the camera. It is relevant to define an algorithm that would determine automatically the position out of the frames. Based on the work of McBride & al. [9], where arc position is given by a weighted mean of the detected luminosity, we developed a software that determines the global position of the arc and the position of one arc root for each electrode by examining the pixels intensity in the vicinity of each runners (green and blue areas in Figure 3). This application also edits each frame of the video for more readability with the size of the chamber, the rotation axis of the contact, the pressure sensor location or other information such as the maximum current like written in red in Figure 4. The calculated positions of the arc are also indicated with crosses (red for the global position of the arc, green and blue for upper and lower arc roots) and electrical waveforms are added on the right of the image.

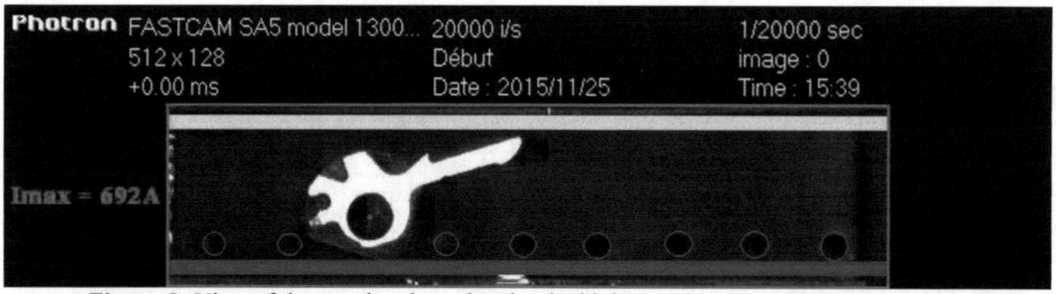

Figure 3: View of the arc chamber taken by the high-speed camera. The green and blue area are the pixel taken into account to determine the positions of respectively the upper and lower arc runner

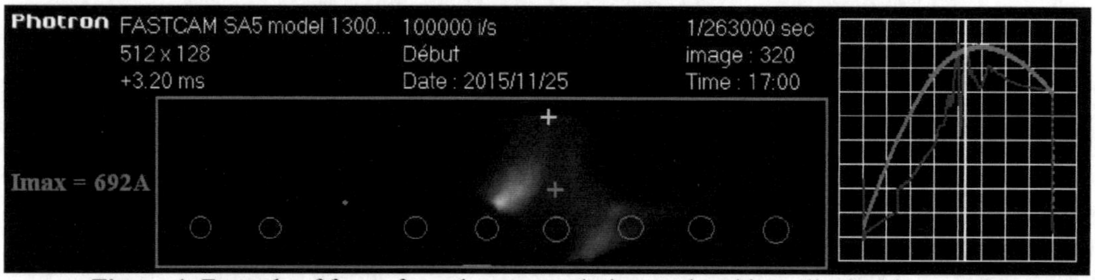

Figure 4: Example of frame from the camera during arc breaking edited with our software. The red and blue curves on the right are the measured current and voltage.

Other post-treatments software has been developed in order to perform statistics on the electrical and pressure measurements. Phenomena like arc ignition and commutation are detected with an algorithm so that the large number of tests can be quickly examined.

3. Influence of gas exhausts

On each extremities of the chamber, two 5mm diameter openings are performed allowing the gas to flow out of the geometry. By closing them we modify the pressure formation and gas flow which impacts the arc movement. Four configurations are considered namely: Open-Open where all exhausts are open; Open-Closed where the downstream exhausts have been shut; Closed-Open where the upstream exhausts are closed and the downstream exhausts are open; Closed-Closed where there are no gas exhausts. For those configurations, the tests were also performed with three levels of intensity with maxima of 631, 1608 and 2852A. The results presented in this section have been obtained with an opening speed of 2.91m/s and a distance between the ignition point and the upstream exhaust of 45 mm.

3.1. High-speed imaging

The global position of the arc determined using the software previously described is shown for three currents in Figures 5-7. We can see a movement which is generally divided in two phases: a slow movement while the arc is bent between the moving contact and the lower arc runner and then a very fast displacement when hot gasses have been spread inside the chamber and the arc commutates.

In the first phase of the movement (between 1 and 2.5ms), the exhaust configuration does not seem to affect the movement of the arc. In the second phase, we see a fast displacement when the arc commutates for Open-Open and Closed-Open configurations. When all exhausts are closed, this movement is slower due to the stagnation of the gas in front of the arc. The worst case is Open-Closed where we can spot that the arc is hardly moving from the contact since the gas fluxes are directed toward the back of the chamber.

Those results are for single tests. Since breaking arcs behaviour is rather chaotic, we will present a statistic analysis on pressure and electrical measurements in the next section.

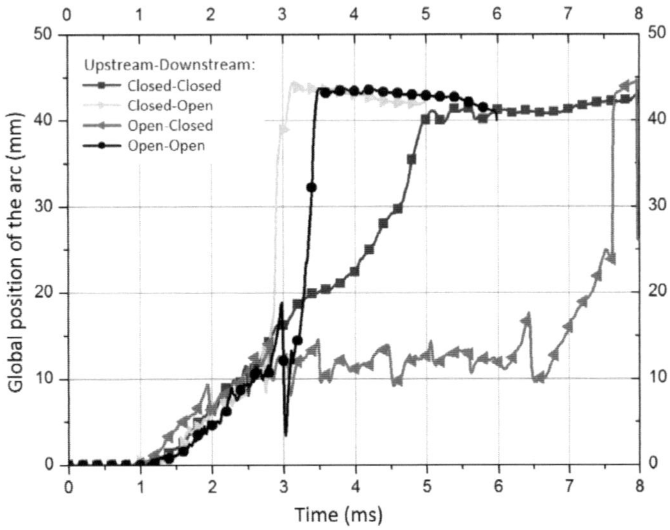

Figure 5: Global position of the arc determined by the method described in §2.5 for different configurations of gas exhaust and a maximal current of 631A.

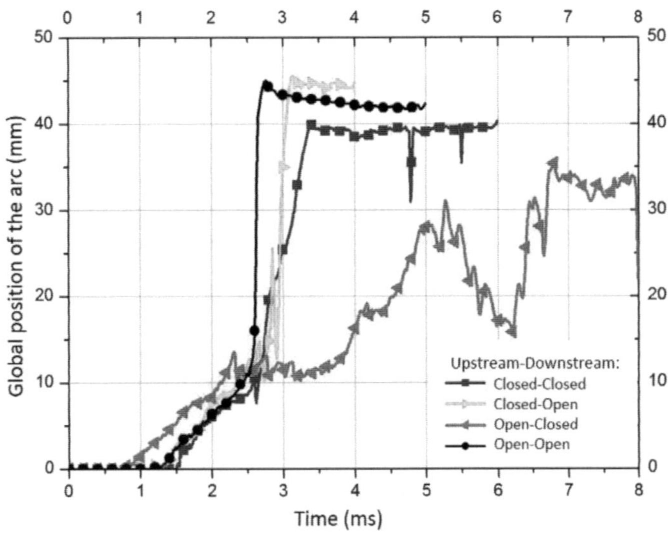

Figure 6: Global position of the arc determined by the method described in §2.5 for different configurations of gas exhaust and a maximal current of 1608A.

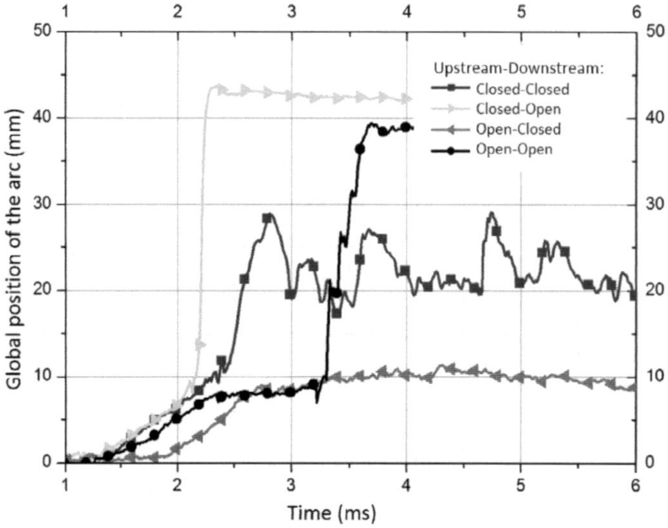

Figure 7: Global position of the arc determined by the method described in §2.5 for different configurations of gas exhaust and a maximal current of 2852A.

3.2. Analysis of the electrical measurements

As seen in Table 1, a higher current level tends to start the commutation process sooner except for the cases where the downstream exhausts are closed. A greater current intensity leads to stronger Laplace forces which bend the arc toward the upper arc runner. But if the downstream exhausts are closed, there will be a pressure elevation in front of the arc preventing it to reach the upper rail. Especially in the Open-Closed configuration, the gas flow is directed to the back of the chamber, acting against the Laplace forces.

Exhausts configuration		Open-Open	Open-Closed	Closed-Open	Closed-Closed
Peak current:	2852 A	1.25 ± 0.09 ms	2.46 ± 0.25 ms	1.41 ± 0.35 ms	1.25 ± 0.27 ms
	1608 A	1.51 ± 0.01 ms	1.24 ± 0.19 ms	1.37 ± 0.06 ms	1.12 ± 0.04 ms
	631 A	1.74 ± 0.06 ms	1.96 ± 0.63 ms	1.80 ± 0.24 ms	1.61 ± 0.03 ms

Table 1: Time between ignition and start of commutation for 3 levels of current

Table 2 shows the commutation duration which is the time between the instant when the contact current gets smaller than the total current and the instant where no current is flowing through the contact. The influence of the intensity is very strong for this topic. At high current, the configuration showing the shorter commutation is the one where all exhausts are open although it has rather long commutation length at low current. On the other hand, the Closed-Open proves to be efficient for smaller current but not for high intensity. For low current, closing the upstream exhaust greatly reduces the commutation time as it should push the arc toward the end of the chamber. Generally, we see that closing some exhausts at high current extends the commutation. The hot gas must leave the chamber in order to prevent restrikes on the mobile contact.

Exhausts configuration		Open-Open	Open-Closed	Closed-Open	Closed-Closed
Peak current:	2852 A	0.78 ± 0.35 ms	6.93 ± 0.27 ms	1.84 ± 0.13 ms	3.53 ± 0.21 ms
	1608 A	1.27 ± 1.58 ms	5.10 ± 0.82 ms	0.79 ± 0.63 ms	3.18 ± 0.27 ms
	631 A	1.81 ± 1.82 ms	5.49 ± 2.23 ms	0.21 ± 0.16 ms	0.27 ± 0.10 ms

Table 2: Commutation duration for 3 levels of current

3.3. Pressure measurements

Gas exhausts reveal a major impact on the pressure level obtained in the chamber. The 10mm chamber is broad enough to allow the gas to flow on the side of the arc and to spread the pressure homogeneously. Thus, all the pressure difference measured in our experiment is within the exhausts, between the inside of the chamber and the free air. As seen in Table 3, pressure is easily doubled when half the exhausts are shut, except for the Closed-Open case at low current where the amount of hot gases is reduced and pushed out of the geometry as the arc moves forward. Pressure is even more strongly increased if the chamber is hermetic. The pressure level is an important parameter for industrial design of the circuit breaker. High pressure often leads to explosion or deterioration of the devices. Also, pressure has an effect on the properties of the medium and could change the arc voltage which is significant for limiter circuit breakers.

Exhausts configuration		Open-Open	Open-Closed	Closed-Open	Closed-Closed
Peak current:	2852 A	0.60 bar	1.51 bar	1.78 bar	3.42 bar
	1608 A	0.46 bar	0.99 bar	0.81 bar	2.82 bar
	631 A	0.17 bar	0.39 bar	0.18 bar	1.43 bar

Table 3: Mean maximal pressure relative to atmospheric pressure for 3 levels of current

4. Influence of upstream volume

If the pressure formation behind the arc impacts the arc behaviour, reducing the upstream volume should also modify the arc motion. In Table 4, it is clear that reducing the volume behind the moving contact starts the commutation earlier by increasing the pressure force toward the end of the chamber. This result is in agreement with the work of McBride & al [4]. However, investigations on commutation duration do not show a significant influence of this parameter. Pressure level is given in Table 5 to depend on of the chamber volume. This is because all the hot gas produced by the electrical arc does not have the time to evacuate the chamber through the narrow exhausts.

Exhausts configuration		Open - Open		Closed - Open	
Distance of upstream wall		30 mm	45 mm	30 mm	45 mm
Peak current:	2852 A	1.26 ± 0.06 ms	1.25 ± 0.09 ms	1.02 ± 0.04 ms	1.41 ± 0.35 ms
	1608 A	1.04 ± 0.10 ms	1.51 ± 0.01 ms	1.08 ± 0.06 ms	1.37 ± 0.06 ms
	631 A	1.29 ± 0.09 ms	1.74 ± 0.06 ms	1.54 ± 0.16 ms	1.80 ± 0.24 ms

Table 4: Time between ignition and start of commutation for 3 levels of current

Exhausts configuration		Open - Open		Closed - Open	
Distance of upstream wall		30 mm	45 mm	30 mm	45 mm
Peak current:	2852 A	0.99 bar	0.60 bar	2.12 bar	1.78 bar
	1608 A	0.51 bar	0.46 bar	1.11 bar	0.81 bar
	631 A	0.25 bar	0.17 bar	0.54 bar	0.18 bar

Table 5: Mean maximal pressure relative to atmospheric pressure for 3 levels of current

5. Conclusion & perspectives

Investigation on arc breaking in a simplified low-voltage circuit breaker chamber has been conducted and gives many clues on arc behaviour that can be transferred to the design of industrial devices. The geometry of the chamber is critical for the fast movement of the arc from the contact region to the splitter plates and an efficient working of the system. The exhaust configuration and volume of the chamber have demonstrated their ability to strongly modify the breaking process.

On the one hand, it is necessary to permit the gas to flow out of the chamber to ensure a good displacement of the arc. On the other hand, the amount of hot material escaping the circuit breaker must be controlled to prevent damages to surrounding apparatuses. Considering also that a given configuration would not work for all level of current, it is clear that developing new breaking solutions is very difficult.

Further parameters may be studied with our set-up. The width of the chamber has an impact on the movement of the arc [6] but also for a narrow geometry the thermal plasma could obstruct the chamber and lead to a pressure gradient that would be detected by our sensors. Correlation between arc movement and local pressure variation should carry interesting information. The material of the walls of the experiment could be changed from plastic to non-ablative ceramics, modifying the pressure formation and the nature of the plasma medium. Moreover, movement of the arc root for electrode of different alloys could be compared. Finally, if we focus on extinction of the arc, splitter plates could be added within the chamber.

In parallel with this experimental work, computational fluid dynamics simulation using finite volume method are performed in similar 3-D configurations for both validation of the model and explanation of the measurements [10]. Such numerical tools would greatly help the development of new products by allowing simulation of a specified architecture before the realisation of a prototype.

References

[1] C.W. Brice, R.A. Dougal & J.L. Hudgins, *Review of Technologies for Current-Limiting Low-Voltage Circuit Breakers*, IEEE trans. indus. app., vol.**32** no.5, 1996

[2] Y. Wu, M. Rong, G. Yang & G. Hu, *Numerical Analysis of the Arc Plasma in a Simplified Low-Voltage Circuit Breaker Chamber with Ferromagnetic Materials*, Plasma sci. & technol., vol.**7** no.4, 2005

[3] M. Lindmayer & M. Springstubbe, *3D-Simulation of arc motion between arc runners including the influence of ferromagnetic material*, Proceedings of the 47th IEEE Holm Conference on Electrical Contacts, 2001

[4] J.W. McBride, K. Pechrach & P.M. Weaver, *Arc Motion and Gas Flow in Current Limiting Circuit Breakers Operating With Low Contact Switching Velocity*, IEEE trans. comp. pack. technol., vol.**25** no.3, 2002

[5] X.Li, D. Chen, Q. Wang & Z. Li, *Simulation of the Effects of Several Factors on Arc Plasma Behavior in Low Voltage Circuit Breaker*, Plasma sci. & technol., vol.**7** no.5, 2005

[6] M. Lindmayer & J. Paulke, *Arc Motion and Pressure Formation in Low Voltage Switchgear*, IEEE trans. comp. pack. manufact. technol., vol.**21** no.1, 1998

[7] K.A. Berger, B. Gessl, W. Schneider & W. Rieder, *Arc Motion and Wave Propagation in Arc Chambers With Lateral Chinks*, IEEE trans. comp. pack. technol., vol.**25** no.3, 2002

[8] P.R. Zeller & W.F. Rieder, *Arc Structure, Arc Motion, and Gas Pressure Between Laterally Enclosed Arc Runners*, IEEE trans. comp. pack. technol., vol.**24** no.3, 2001

[9] J.W. McBride, P.M. Weaver & P.A. Jeffery, *Arc Root Mobility During Contact Opening at High Current*, IEEE trans. comp. pack. manufact. technol., vol.**21** no.1, 1998

[10] J. Quéméneur, P. Freton, M. Masquère, J.-J. Gonzalez, P. Joyeux, *Cathode Arc Root Movement : Models Comparison*, Plasma phys. & technol., vol.**2** no.2, 2015

IOP Conf. Series: Journal of Physics: Conf. Series **825** (2017) 012013 doi:10.1088/1742-6596/825/1/012013

Investigation of the AC Plasma Torch Working Conditions for the Plasma Chemical Applications

A A Safronov, O B Vasilieva, J D Dudnik, V E Kuznetsov*, V N Shiryaev, D I Subbotin and A V Pavlov

Institute for Electrophysics and Electric Power of Russian Academy of Sciences (IEE RAS), Dvortsovaya emb. 18, 191186, St.-Petersburg, Russia

*kuznetsov@iperas.nw.ru

Abstract. The presented design and parameters of a three-phase AC plasma torch with the power up to 500 kW, flow rate of air 30–50 g/s (temperature up to 5000 K) could be used in different plasma chemical processes. Range of measured plasma temperature is 3500–5000 K. The paper presents investigations of the plasma torch operation modes for its application in plasma chemical technologies. Plasma chemical technologies for various purposes (processing, destruction of various wastes, including technological and hazardous waste, conversion or production of chemicals to obtain nanoscale materials, etc.) are very promising in terms of the process efficiency. Their industrial use is difficult due to the lack of inexpensive and reliable plasma torches providing the desired level of temperature, enthalpy of the working gas and other necessary conditions for the process. This problem can be solved using a considered design of a three-phase alternating current plasma torch with power of 150–500 kW with working gas flow rate of 30–50 g/s with mass average temperature up to 5000K on the basis of which an industrial plasma chemical plant can be created. The basis of the plasma torch operation is a railgun effect that is the principle of arc movement in the field of its own current field. Thanks to single supply of power to the arc, arcs forming in the discharge chamber of the plasma torch move along the electrodes under the action of electrodynamic forces resulting from the interaction of the arc current with its own magnetic field. Under the condition of the three-phase supply voltage, arc transits from the electrode to the electrode with change in the anodic and cathodic phases with frequency of 300 Hz. A special feature of this design is the ability to organize the movement of the arc attachment along the electrode, thus ensuring an even distribution of the thermal load and thus achieve long time of continuous operation of the plasma torch. The parameters of the plasma jet of the plasma torch and the single-phase three-phase plasma injector for use in a plasma-chemical unit for production of nano-dispersed materials are described in the paper.

IOP Conf. Series: Journal of Physics: Conf. Series **825** (2017) 012013 doi:10.1088/1742-6596/825/1/012013

1. Introduction

Plasma torches usage in plasma-chemical technologies most often refers to such areas as plasma processing of hazardous waste (medical, biological, toxic, etc.) by plasma pyrolysis and gasification, vitrification of slag, processing of associated gas with production of synthesis gas suitable as a raw material for electricity generation or liquid fuel production, plasma-chemical synthesis of ultrafine powders [1]. The latter may be of considerable interest because introduction of small (a few percent) amounts of nanopowders in the material leads to significant improvement of operational properties of the alloys [2].

The most widely used oxides are respectively SiO_2, $A1_2O_3$ and TiO_2 [3].The main prerequisites for the use of ultrafine powders are small particle size commensurate with the size of nucleation, high sedimentation stability of the particles in the melt and the ability to achieve a high number concentration of particles at low mass concentration. Application of ultrafine powders of refractory compounds as modifiers of metals and alloys makes it possible to control the structure of the castings in the a-priori intended direction allowing influence on grain size and shape, distribution and amount of carbide, intermetallic and other phases. Technology modification of ultrafine powders can increase both strength and ductility of cast metal. Substantial change in alloy structure resulting from the modification results in increased operating characteristics: limit strength increases by 10–30%, ductility increases in 1.5–2 times, endurance limit increases by 25–40%. Modifying effects were tested on aluminum and nickel, and magnesium alloys, high-chrome cast iron, corrosion-resistant steels [4,5,6].

A feature of the plasma chemical process for production of ultrafine oxide materials is offered for consideration and implemented on the basis of AC plasma torch with rail electrodes operating in a wide range of powers and flow rates of the oxidizing gas with the possibility of producing electrodes from a variety of materials such as aluminum, and capable of providing high speed heating to the temperature of about 5000K and cooling of the processing material, which makes it particularly attractive for use in the plasma chemical technologies for production of ultrafine materials.

2. Description of the experimental installation

Photo of the working three-phase plasma torch is shown in Figure 1.

Figure 1. Photo of AC plasma torch with rail electrodes [7]

The principle of electrodynamic arcs movement in a field of its own current (railgun effect) is the basis of its operation. Quick movement of the arc spot along the electrode under the action of the electrodynamic and gas dynamic forces distributes the thermal load from spot along the length of the electrode that enables the use of water-cooled electrodes made of relatively low-melting material with high heat conductivity and capable of operating in oxidizing environments.

The structure of the plasma torch includes the injector which is a single-phase high-voltage plasma torch of low power. The injector creates in the minimum electrode gap the plasma stream with $n_e = 10^{13}–10^{14}$ cm^{-3} sufficient to ignite the main arcs and smooth transition of current through zero. Arcs are initiated between the main electrodes in the zone of minimum electrode gap, and then arc attachments move along the arc surfaces of diverge electrodes.

Two arcs burn simultaneously in the discharge chamber of a single chamber three-phase plasma torch with rail type electrodes and they fill the available space.

The dimensions of the three-phase plasma torch: L – 60 cm, D – 45 cm, diameter of the outlet nozzle 6 cm, minimal distance between rail electrodes is 2 cm, length of rail electrode is 25 cm.

Features of this plasma torch design allow operation in a wide range of power settings and working gas enthalpy Figures 2 and 3.

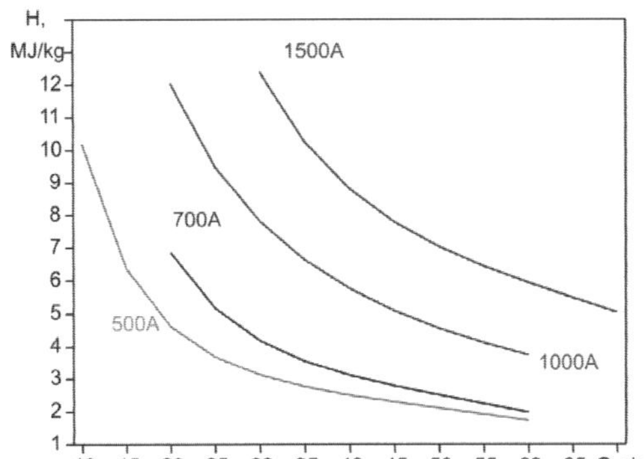

Figure 2. Dependence of the heat content of the gas from air flow rate at atmospheric pressure. The curves are marked in accordance with the values of the current strength of the electric arc.

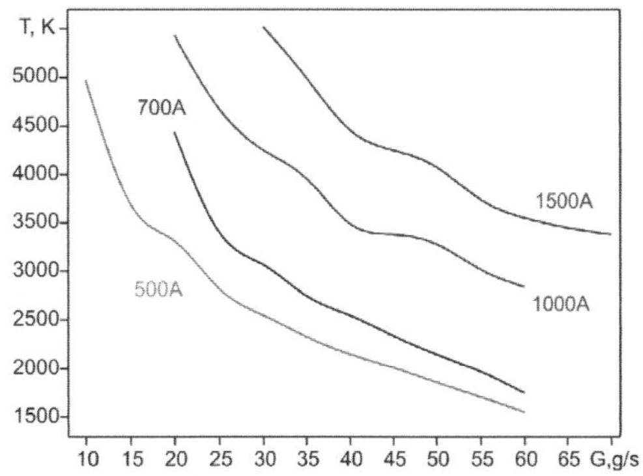

Figure 3. Dependence of average mass temperature of the working gas from air flow rate at atmospheric pressure. The curves are marked in accordance with the values of the current strength of the electric arc [4].

Figure 3 shows the dependence of average mass temperature of the working gas from air flow rate. This temperature is determined by the balance of electric energy and the energy that is lost with cooling system. It was found that the arc voltage drop depends slightly on the air flow rate [7]. It means that the electric power of the described plasma torch with the fixed current depends slightly on the air flow rate. So the enthalpy and average mass temperature of the working gas decrease with the air flow rate increase.

Longitudinal and radial temperature profiles in the flame of the plasma torch were obtained using spectral measurements (Figure 4). Figure 4 is a schematic diagram of spectral measurements of the radiation plume of a low-voltage AC arc plasma torch.

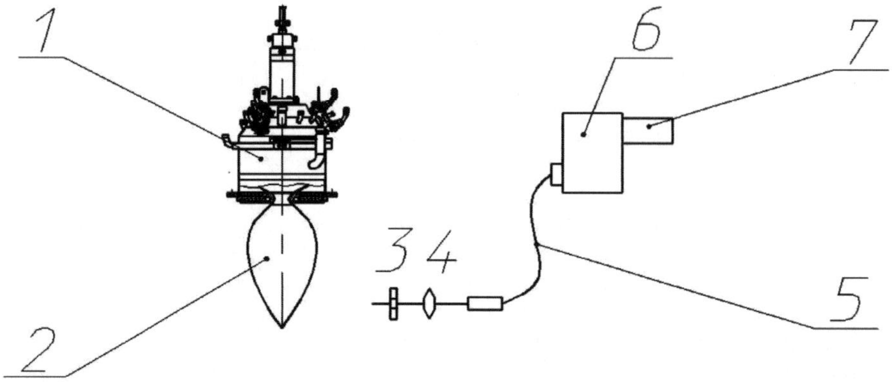

Figure 4. Scheme of spectral measurements of the radiation plume of the AC plasma torch with rail electrodes: 1 – plasma torch; 2 – plume; 3 – diaphragm; 4 – lens with focal length 22 cm; 5 – optical fiber; 6 – spectrometer Andor Shamrock SR–303i–A; 7 – CCD camera Andor 420–DU–FK.

The plasma jet at the outlet of the nozzle of the three-phase AC plasma torch is an axisymmetric cone flow (Figure 1). Temperature measurements in the flame of the AC plasma torch with rail electrodes (Figure 5 and 6) are made for different modes. The results of the measurements are presented in Table 1.

The method of the relative intensities in the approximation of local thermodynamic equilibrium was used to determine the electron temperature [8]. The torch is formed outside from the electric arc for this type of plasma generator. So there are no any electric or magnetic fields in the torch area that can lead to the non-equilibrium plasma condition. Moreover, LTE is typical for the air plasma in electric arc at atmospheric pressure [9].

The relative intensities are calculated for four pairs of copper atom's spectral lines (in nm): CuI 515.32 and CuI 510.5, CuI 521.82 and CuI 510.5, CuI 515.32 and CuI 570.02, CuI 521.82 and CuI 570.02. These temperatures are comparable to the margin of statistical error, which does not exceed 10%.

Radial symmetry of the torch is verified by measurements of the emission spectra of the copper atoms from different angles throughout the length of the cross section of the torch. So it is possible to use the Abel transform to obtain radial temperature profiles [10].

Table 1. Parameters of a three-phase AC air plasma torch at atmospheric pressure.

Parameter	Mode of operation of the power supply		
	1	2	3
Short-circuit current of power supply, A	500	700	1000
Voltage of the supply network, V	480	480	480
Flow rate of plasma forming gas, g/s	30	30	25
Arc current (root-mean-square active value), A	405	560	840
Arc voltage (root-mean-square value), V	240	250	230
Power no less than, kW	125	195	295
Specific electrode erosion, g/C	$6.5 10^{-6}$	$8.5 10^{-6}$	$8.2 10^{-6}$
Life time of continuous operation without electrode replacement, hour	More than 100	85	35

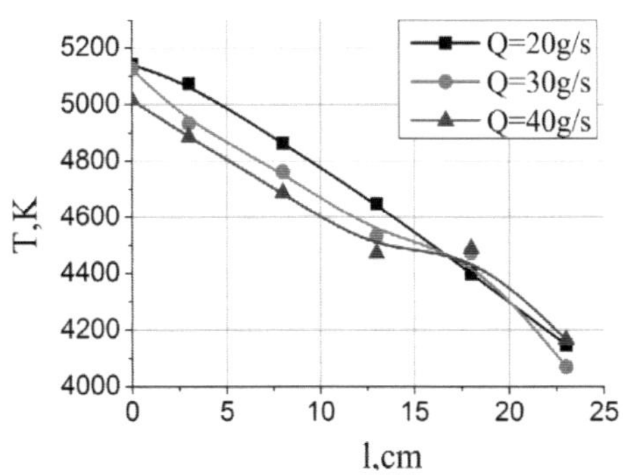

Figure 5. The longitudinal profiles of temperature in the flame center of the plasma torch with rail electrodes at root-mean-square arc current value of 500 A, and different working gas flow rates Q: l – distance from the nozzle outlet.

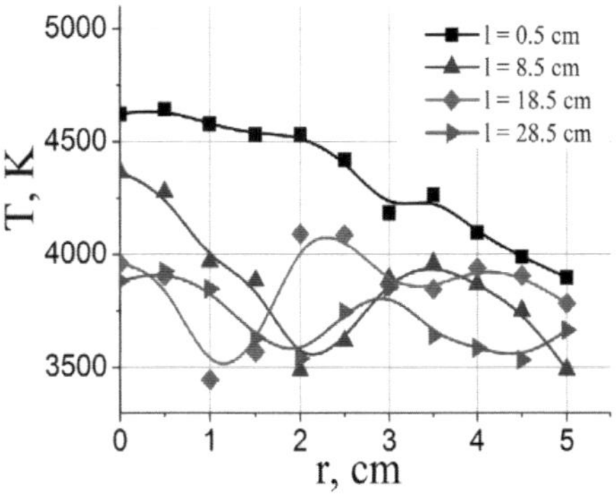

Figure 6. Radial temperature profiles in different sections of the flame of the arc plasma torch with rail electrodes at the root-mean-square current value of the arc current of 500 A, and working gas flow rate Q = 30 g/s: l – distance from the nozzle outlet in cm, T – temperature, r – radius.

It can be concluded from Figure 5 that the recorded maximum temperature at the nozzle outlet of the plasma torch is practically unchanged at different flow rates.

As can be seen from Figure 3 at flow rate of the working gas up to 25 g/s the average mass temperature of the plasma jet even at the minimum power is above 2500K, which is also acknowledged by the spectral measurements Figure 5 and Figure 6. As long as the statistical error of temperature measurements was 10 %, the temperature fluctuations along the radius can be attributed to the measurement error.

Therefore, in addition to "traditional" use of plasma torches as a source of high-enthalpy gas flows in waste processing plants or in conversion plasma-chemical technologies aimed, for example, on the production of synthesis gas the use of the plasma torch can be substantially expanded.

High thermodynamic characteristics allow usage of this design as a part of the plasma torch plasma chemical plant for production of ultrafine materials Figure 7.

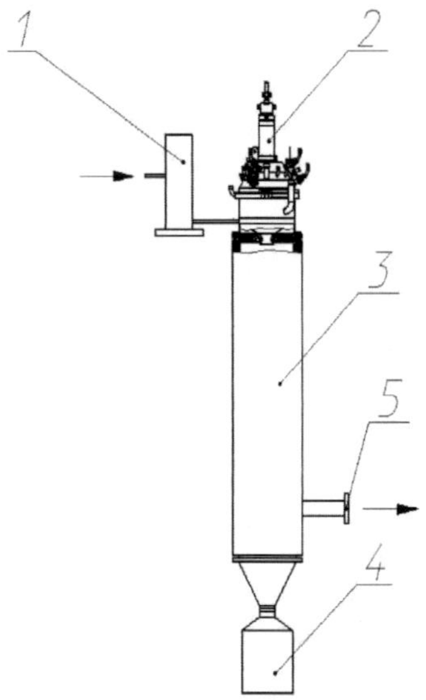

Figure 7. Plasma-chemical plant on the basis of a three-phase AC plasma torch with rail electrodes. 1 – supply system of the processed material, 2 – plasma torch, 3 – plasma reactor 4 – product collection system, 5 – gas outlet

Plasma torch design allows organization of the processed powder introduction at several points relatively to the plasma stream, including directly into the electric arc chamber. The size of the cylindrical water-cooled plasma chemical reactor: diameter 300 mm and height 1350 mm. In this case the plasma torch body is used as the first step of the plasma chemical reactor.

3. Application

The design features and the ability to ensure high plasma jet temperatures even at the minimum power of the plasma torch (about 100 kW) permits the plasma torch application as a part of the installation for plasma chemical synthesis of ultrafine oxide powders (Al_2O_3, ZrO_2, MgO, SiO_2). One of the ways of nanoparticles formation is metal vapour condensation at quenching of high temperature flow.

Thermodynamic estimaton of conversion of precursor compounds into ultrafine oxide powders ZrO_2, Al_2O_3, MgO was carried out for initial evaluation of the applicability of the plasma torch. The computation was performed using the program "Chemical WorkBench" ver.3.5, Kinetic Technologies Ltd. [11], based on searching of minimum of Gibbs free energy. Precursors: ZrO_2 – $ZrCl_4$ (zirconium (IV) chloride); Al_2O_3 – Al_2O_3 (aluminum oxide); MgO – Mg (metal magnesium).

Synthesis of zirconia

Zirconium oxide is required for the production of refractory materials, high-melting glasses, various kinds of ceramic, thermal barrier coatings and reaction proceeds according to the equation:

$$ZrCl_4+O_2=ZrO_2+2Cl_2 \quad (1)$$

Initial data for calculation: air specific flow rate of is 0.65 kg/kg of zirconium chloride; temperature range is 500–5000K. Specific flow rate shows air flow rate required for the reaction of 0.65 kg of air for treatment of 1 kg of substance.

Minimum temperature for zirconia production from zirconium chloride by the reaction (1) is 600 °C. Thus, at 340 °C zirconium chloride sublimes. Production in crucible depends on the composition of the product mixture from the temperature as it is shown in Figure 8.

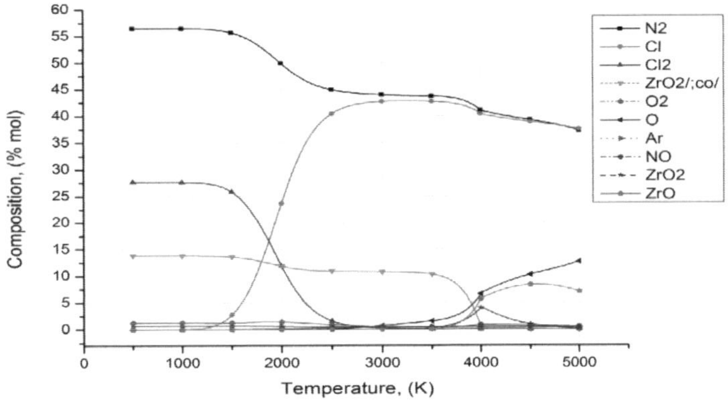

Figure 8. Composition of the product mixture at heating of zirconium chloride with air

Temperature of phase transition (evaporation) of zirconium oxides is about 4000K. In this case, the mechanism of formation of ultrafine structures will be determined by different parameters, in the first place by the temperature of condensation processes. Prior to this, the main mode of the process development is thermal destruction of the original grain.

Synthesis of alumina

Aluminum oxide is used in a wide range of technologies: for optical and structural ceramics, catalysts) etc. Ultrafine alumina is advisable to produce from the coarse grain, so the main task is to achieve a high temperature for the thermal destruction of grain and evaporation. Figure 9 shows the thermodynamic parameters of the alumina heating to a temperature above the melting temperature at the plasma gas flow rate of 1 kg/kg of aluminum oxide. Apparently, the required value of plasma enthalpy for this process is 5 MJ/kg of aluminum oxide.

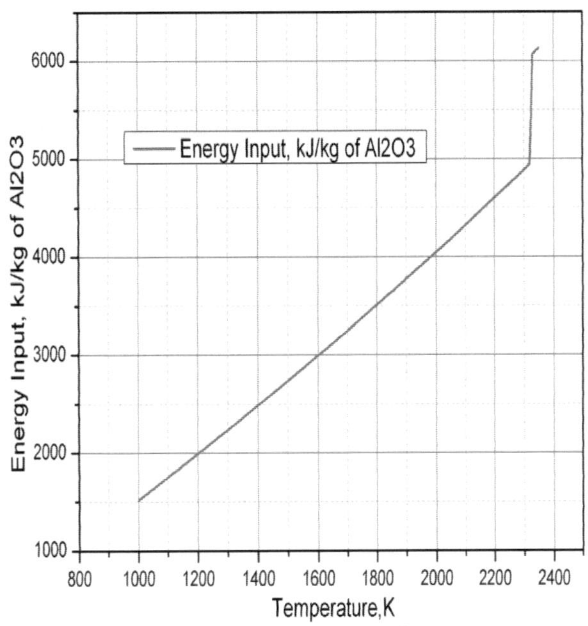

Figure 9. Heating by air plasma

IOP Conf. Series: Journal of Physics: Conf. Series **825** (2017) 012013 doi:10.1088/1742-6596/825/1/012013

Synthesis of magnesium oxide

Magnesium oxide is widely used in mechanical engineering, petroleum industry, manufacture of refractory materials and production of certain types of cement.

The process evaluation was carried out of the equation:

$$2Mg + O_2 = 2MgO \qquad (2)$$

Specific air consumption – 2.88 kg/kg, reaction is exothermic, so the cost of ultrafine powders will be less. Transition to the gaseous phase of magnesium oxide starts at 3000 K (Figure 10). Simultaneously the formation of magnesium metal starts again.

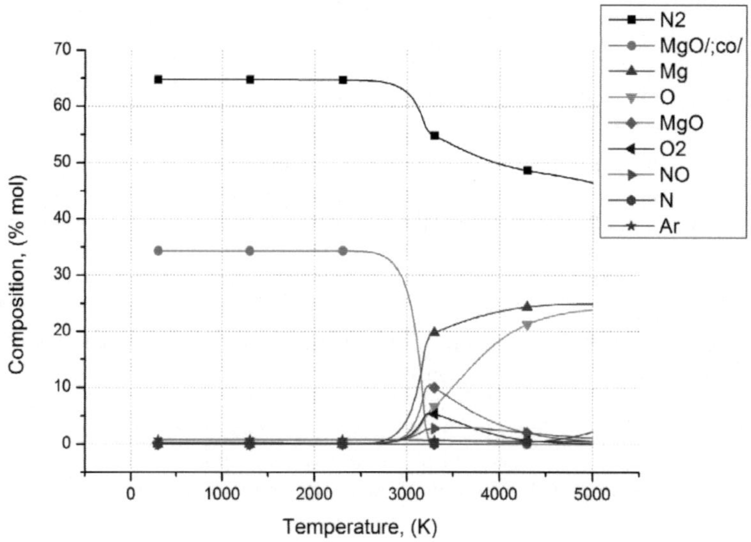

Figure 10. Composition of product mixture at heating of metal magnesium with air.

The temperature of the beginning of the formation of zirconium oxide is 3500 K, alumina – 3500 K, magnesium oxide – 3200 K. Taking into account that the average mass temperature of the working gas for flow rates up to 25 g/s is in the range from 3000 K to 5000 K, and the measured temperature does not fall below 3500K, conditions for the synthesis of these oxides may be created.

Based on the data from [12] it is possible to obtain nanoparticles with size ranging from 10 to 150 nm using the plasma-arc method.

4. Conclusion

It was found during experiments that the high temperature jet area is an ellipsoidal region with dimensions no less than 250x60 mm. The average mass temperature value in this zone is no less than 3000 K. This allows the use of the plasma torch for plasma chemical synthesis of ultrafine oxide powders (Al_2O_3, ZrO_2, MgO, etc.).

References

[1] Trelles J P, Chazelas C, Vardelle A and Heberlein J V R 2009 Journal of Thermal Spray Technology **18** 728–752
[2] Rempel A A 2007 Russian Chemical Reviews **76(5)** 435–461
[3] Malanowski N, Heimer T, Luther W and Werner M 2006 Growth Market Nanotechnology: An Analysis of Technology and Innovation (John Wiley & Sons, Incorporated)
[4] Mändl S and Gerlach J W, Rauschenbach B 2005 Surface and Coatings Technology **196(1–3)** 293–297
[5] Prysiazhnyi V, Svoboda T, Dvořák M and Klíma M 2012 Surface and Coatings Technology **206(19–20)** 4140–4145
[6] Jingxin Yang, Fu-zhai Cui, In Seop. Lee and Xiumei Wang 2010 Surface and Coatings Technology **205(1)** 182–187
[7] Rutberg Ph G, Safronov A A, Popov S D, Surov A V and Nakonechny G V 2005 Plasma Physics and Controlled Fusion **47** 1681–1696
[8] Griem H 1997 Principles of Plasma Spectroscopy (Cambridge University Press)
[9] Raizer Yu 1997 Gas Discharge Physics (Springer, New York)
[10] Hutchinson I 2002 Principles of Plasma Diagnostics (Cambridge University Press)
[11] Chemical WorkBench, Kinetic Technologies Ltd. Available at: http://www.kintechlab.com/products/chemical-workbench/ (accessed 11 January 2017)
[12] Fadeev A A, Samokhin A V, Alekseev N V and Tsvetkov Y V 2013 Solid State Physics Bulletin of the N.I. Lobachevsky Nizhny Novgorod University **2(2)** 66–71

Analysis of Physics Processes in the AC Plasma Torch Discharge under High Pressure

A A Safronov, O B Vasilieva, J D Dudnik, V E Kuznetsov*, J A Kuchina, V N Shiryaev and A V Pavlov

Institute for Electrophysics and Electric Power of Russian Academy of Sciences (IEE RAS), Dvortsovaya emb. 18, 191186, St.-Petersburg, Russia

*kuznetsov@iperas.nw.ru

Abstract. The paper is devoted to investigation of electrophysical processes in the electric discharge generated by a three-phase AC plasma torch when using a high pressure inert working gas. AC plasma torch design with end electrodes intended for work on inert gases at pressures up to 81 bar is studied. Current-voltage characteristics for different gas flow rates and pressures are presented. Physical processes characteristics of the arising voltage ripples which depend on various working parameters of the plasma torch have been investigated. Arc burning processes in the electric discharge chamber of the three-phase AC plasma torch at various working parameters were photographed.

As it is known, pressure is one of the main factors influencing the speed of chemical reactions. The same can be said about plasma-chemical reactions [1]. For example, it was found that the optimal conditions for the synthesis of NO in thermal electric arc plasma lie at temperature of 3000-3500 K and pressure of about 20 bar [2].

In addition, high pressure in the plasma chemical reactor allows quenching of the reaction products in the Laval nozzle, which greatly simplifies synthesis installation structure [3].

To date, electric arc plasma torches are the most suitable heat source for large-scale plasma-chemical plants from all of thermal plasma generating units [4].

Direct current systems are the most widely used, while electric arc AC installations are still exotic. However, due to the periodic change of polarity at the electrodes, they have a higher life time [5]. AC plasma torches using air as a working gas have been developed in IEE RAS. These kinds of systems are able to continuously operate to up to 2000 h. However development of AC plasma torches using higher pressures and understanding of the physical processes within those systems remain highly relevant [6-8].

Experimental setup

Figure 1 shows a schematics and a photograph of the electric arc AC plasma torch designed to operate at a high pressure of the working gas. Arc chamber 1 and three electrodes 4 are water cooled, the chamber is grounded. Compensation of thermal stresses appearing at the expense of arc chamber heating is realized by corrugated hose 2. Spiral guides 8 organizing the flow of coolant are arranged inside the cooling jacket 3.

IOP Conf. Series: Journal of Physics: Conf. Series **825** (2017) 012014 doi:10.1088/1742-6596/825/1/012014

The working gas is supplied through the branch pipe 5 into the vortex chamber. The gas is fed inside the arc chamber through a series of openings arranged tangentially to its inner surface.

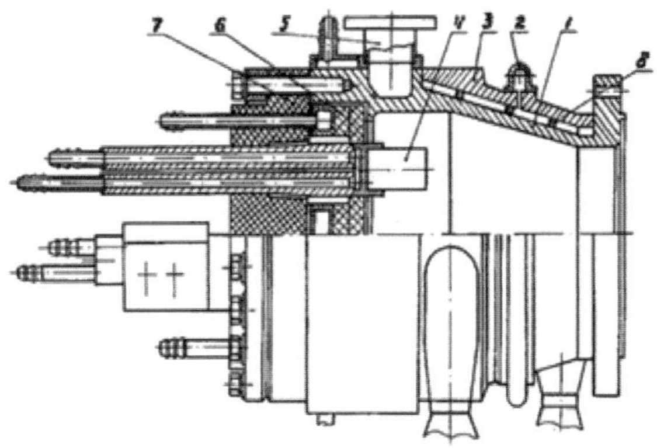

Figure.1 (a) Schematics of three-phase electric arc plasma torch: 1- arc chamber; 2- corrugated hose; 3- cooling jacket; 4- electrodes; 5- branch pipe; 6- insulators; 7- insulators; 8- spiral guides.

Figure 1(b) Photograph of three-phase electric arc plasma torch.

The light yellow elements are used for connection of power supply cable.

The electrode 4 is a copper body where a set of tungsten rods with diameter of 16 mm is installed. Electrode holders have channels for cooling water supply. Insulators 6 and 7 are used for their galvanic isolation from the chamber. Load voltage of the power supply is 800 V, short circuit current of the power supply varies from 300 A to 500 A.

Figure 2 shows the current-voltage characteristics of the plasma torch at different pressures.

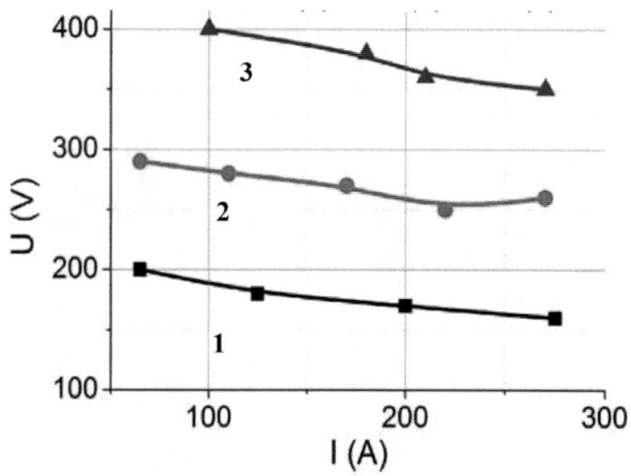

Figure 2. Current-voltage characteristics of the plasma torch at different pressures of the working gas in the chambers: 1 working gas flow rate 24 g/s, pressure 1 bar; 2 working gas flow rate 24 g/s, pressure 14 bar; 3 working gas flow rate 32 g/s, pressure 30 bar.

Optical scheme of high speed shooting is shown in Figure 3.

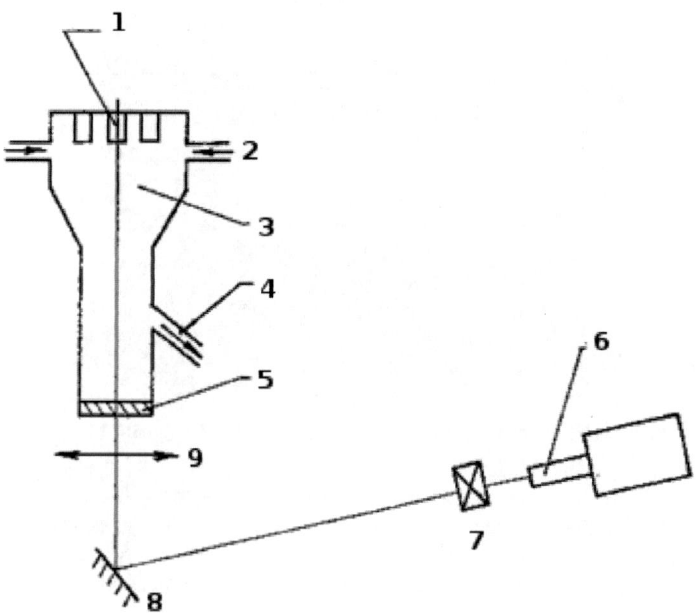

Figure 3. Optical scheme of shooting: 1 – electrodes: 2- gas; 3- plasma torch chamber; 4- exhaust gas; 5-quartz glass; 6-high speed camera; 7- shatter; 8- mirror; 9- lens

Single frame shooting was synchronized with the voltage on the arcs. Arc voltages and linear currents were recorded on the cathode oscilloscope. Light signals corresponding to the arc voltages were recorded on the film. The form of the arc voltages is decomposed in Fourier series, and RMS was calculated on the first harmonic. Linear currents are nearly sinusoidal. Current-voltage characteristics were recorded at the transition from the three-phase to single phase.

Physical processes in the plasma torch chamber were investigated using high-speed shooting at frequency of 3600 frames/s. Exposure time is 0.03 ms, period of power network is 20 ms. The speed of shooting does not allow seeing the commutation of the arc between the electrodes. Nitrogen was used as a working gas during the investigation of the described installation. The pressure in the arc chamber was varied in the range 1 – 30 bar.

Shooting was carried out at different modes and the most characteristic ones were included into the paper.

The analysis of experimental data showed the decisive role of gas-dynamic processes in the formation of a conductive channel of the arc and gas heating. We observed strong influence the gas flow rate on the arc shape and its emission characteristics, as well as on its stability. Furthermore, arc stability to a large degree depends on the current magnitude. Thus, it was found that at pressure of 1 bar and at arc current strength of 220 A the upper limit of the working gas flow rate at which a stable arc can be observed lies at 25 g/s.

The power supply has the output voltage 800V, short circuit current can vary from 300 A to 500 A. Adjustable inductance switched in each phase has been used.

Video frames from the plasma torch chamber were compared with different pressure therein during the investigation. Figure 4 shows the frames of reswitching of arcs from one electrode pair to another at atmospheric pressure. It can be seen that it is possible to identify both conducting channel, and convection flows of heated gas. As it is seen in the main number of frames the conductive channel is not seen. We can see only separate areas of heated up gas having non-uniform form. In the middle frame the conductive channel of several mm width can be seen.

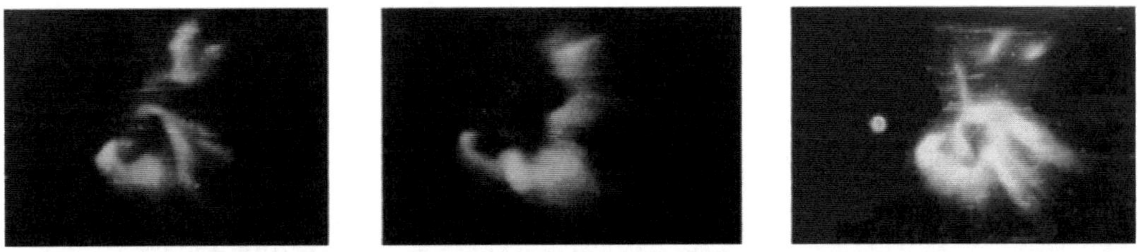

Figure 4. Frames of reswitching of arcs in the plasma torch chamber at atmospheric pressure (root mean square value of arc current – 200A, working gas flow rate – 24 g/s)

On the other hand, by increasing the pressure the arc channel radiation in the visible region disappears, there is only gas glow (Figure 5)

Figure 5. Frames of reswitching of arcs in the plasma torch chamber at pressure of 14 bar (root mean square active value of arc current – 260A, working gas flow rate – 24 g/s)

In Figure 5 it can be seen that the discharge current in the chamber does not live any traces, but only the glowing of near-electrode areas. It is necessary to have spectral measurement data for interpretation of the observed picture.

Based on studies of the three-phase AC plasma torch with nitrogen as a working gas it has been found that by increasing the pressure inside the chamber: the current-voltage characteristic has falling character. This can be explained by the fact that in the range of operating currents of the plasma torch the conductivity increases; arcs radiation disappears in the visible range; glowing of convective gas flows intensifies around the arcs.

Voltage increases at pressure increasing. This is due to the increase of strength on the arc and is likely due to contraction of the electric column. That is, the arc column diameter reduces and its temperature increases. That leads to increase in the arc voltage, but the average mass temperature of the gas decreases.

Absence of arcs glow in the visible range can be explained by a lower degree of dissociation of nitrogen molecules that converts radiation into ultraviolet range.

References

[1] Fridman A 2008 Plasma Chemistry (New York: Cambridge University Press) p 978
[2] Polak L S 1965 Kinetics and thermodynamics of chemical reactions in a low-temperature plasma (Moscow: Nauka) p 255
[3] Polak L S, Ovsiannikov A A, Slovetsky D I and Vurzel F B 1975 Theoretical and Applied Plasma Chemistry (Moscow: Nauka)
[4] Fabry F, Rehmet C, Rohani V and Fulcheri L 2013 Waste and Biomass Valorization **4** 421–439
[5] Fulcheri L, Fabry F, Takali S and Rohani V 2015 Plasma Chemistry and Plasma Processing **35** 565–585
[6] Bogomaz A A, Budin A V, Kolikov V A, Pinchuk M E, Pozubenkov A A and Rutberg Ph G 2003 Doklady Physics **48(1)** 1-4
[7] Budin A V, Pinchuk M E, Kuznetsov V E and F.G. Rutberg 2014 Technical Physics Letters **40(12)** 1061–1064
[8] Pinchuk M E, Bogomaz A A, Budin A V and Rutberg P G 2014 IEEE Transactions on Plasma Science **42(10)** 2434-2435

Steam, methane and carbon dioxide thermal plasma interaction with perhalocarbons

A V Surov[1, 2], S D Popov[1, 2], V E Popov[1], D I Subbotin[1], N V Obraztsov[1], J A Kuchina[1], E O Serba[1], Gh V Nakonechny[1], V A Spodobin[1], A V Pavlov[1] and A V Nikonov[1]

[1] Institute for Electrophysics and Electric Power of Russian Academy of Sciences (IEE RAS), Dvortsovaya emb. 18, 191186, St.Petersburg, Russia
[2] Peter the Great St. Petersburg Polytechnical University, Polytechnicheskaya 29,195251, St.Petersburg, Russia

E-mail: alex_surov@mail.ru

Abstract. Many perhalogenated hydrocarbons (including carbon tetrafluoride and carbon tetrachloride) are limited in production, because they have negative impact on the environment. A number of plasma technologies for perhalogenated hydrocarbons processing is proposed in addition to chemical methods. Usage of thermal plasma of methane and steam may be effective for the processing of these materials into carbon monoxide and certain hydrogen halide. The first stage of the investigation shows that hydrogen chloride is the only chlorine-containing compound during carbon tetrachloride processing.

1. Introduction

Chlorinated organic compounds are widely used in industry as intermediates and end products, such as solvents, electric liquids, monomers for the production of plastics, etc. At the moment, the accumulated reserves of high-boiling mixtures of organochlorine compounds require environmentally safe disposal because chlorinated hydrocarbons are the main polluters of groundwater and are actively accumulate in the tissues of living organisms.

In particular, carbon tetrachloride (CCl_4) is one of the most dangerous of all chlorinated hydrocarbon which has a damage effect on the environment and human [1 - 3].

One method of treatment of hazardous organic compounds is incineration. Methods of conventional high-temperature destruction or oxidation of carbon tetrachloride are well studied [4 - 8]. However, the conditions of organic substances combustion cannot ensure their complete destruction and in the exhaust gases may be present dangerous products of chemical underfiring, such as phosgene, dibenzofuran, dioxins, aromatic hydrocarbons (primarily fused aromatics), and other thermally stable products. A number of plasma methods are proposed in addition to chemical ones: microwave plasma, electric arc plasma, non-equilibrium plasma of corona and barrier discharges. Most of the investigations are carried out under oxidizing conditions [9 - 10], which inevitably lead to the formation of toxic gases as reaction products such as phosgene.

A universal plasma heater based on the principle of a rotating arc is described [11]. Almost all organochlorines are decomposed at temperature of 1000-1750 K (with efficiency of > 99.8 %). Under pyrolysis of CCl_4 and $C_{12}H_8Cl_2$ soot was formed, which is unacceptable, since soot is deposited on the electrodes and insulators. The most favorable conditions were found with the addition of H_2O to decompose chlorinated waste. In this case, no soot, no toxic gas Cl_2 were formed.

The team of scientists of the Physics Department of the Spanish University of Cordoba considered the use of microwave plasma to decompose CCl_4 and C_2HCl_3 [12]. A number of byproducts C_2HCl_3, CCl_4, and C_2Cl_4 generated during destruction under optimal conditions are within the range of ppb. Other by-products found in the outgoing gases (CO_2, N_2O and NO) may be captured by suitable filters.

It was found that CCl_4 conversion increases from 85% to 90% with an increase in steam consumption on a laboratory installation with gliding arcs using steam and air [13]. Experiments on chlorinated organic compounds decomposition were carried out at the discharge power of 600 VA and flowrate of 200 l/h. The main products of methyl chloride decomposition were molecular chlorine (Cl_2), trichloromethane ($CHCl_3$) and HCl. In addition, processing of carbon tetrachloride was investigated to obtain chlorinated polymer [14] via a double dielectric barrier discharge reactor. Methane was fed in the reaction to provide a source of hydrogen and carbon for the reaction with CCl_4 and its decomposition products. At temperature of 117 °C and applied voltage of 16 kV, the conversion of CH_4 and CCl_4 was 19.1 % and 30.6 % respectively, while at 157 °C and voltage of 18 kV its was 37% and 44.5 %.

The paper presents a method for decomposition of perhalogenated hydrocarbons by the AC plasma torch operating on mixture of steam, methane and carbon dioxide for plasma reforming of methane [15].

Usage of steam as plasma forming gas instead of air offers several advantages:

- Nitrogen is excluded from the reaction and consequently there is no risk of formation of toxic nitrogen compounds.
- The volume of exhaust gases is 2-3 times lower than with air plasma, which reduces the cost of gas treatment.
- Steam is a high-performance agent preventing soot formation.
- If the initial material contains insufficient amount of hydrogen for chlorine fixation, additional amount of hydrogen produced by reaction of steam with methane can be fed into the high temperature zone.

2. Thermodynamic computation of decomposition of halogenated organic compounds

AC plasma torch with power 120 kW and thermal efficiency of no less than 90 % was considered at simulation of carbon tetrachloride and carbon tetrafluoride decomposition [16]. Flowrates of plasma-forming substances: CO_2 is 2.9 g/s, H_2O is 2.9 g/s, CH_4 is variable. Stoichiometric reaction of carbon tetrachloride and carbon tetrafluoride with steam, carbon dioxide and methane for hydrogen production, needed for complete chlorine fixation into hydrogen chloride was accepted as initial conditions:

$$CCl_4+1.049H_2O+0.427CO_2+0,476CH_4=1.903CO+4HCl \quad (1)$$

Process parameters computation was performed using software ChemicalWorkbench 3.5. Computation conditions: temperature 1500 K (at this temperature methane is completely decomposed under the action of steam [15]), pressure 101325 Pa. Stream 1: H_2O is 41.61% mass, CO_2 is 41.61% mass, CH_4 is 16.78% mass. Stream 2: carbon tetrachloride or carbon tetrafluoride. Composition of the product mixture as a function of Stream 1 for tetrachloromethane and tetrafluoromethane is presented in figures 1 and 2.

As can be seen if flow rate of steam, carbon dioxide and methane mixture is more than 0.3 kg/kg of carbon tetrachloride and 0.5 kg/kg of carbon tetrafluoride the products have no solid phase (graphite) and the only halogen-containing compound is certain hydrogen halide.

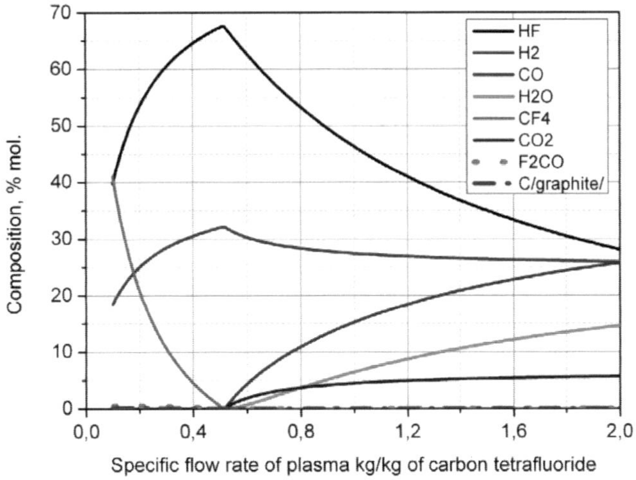

Figure 1. Decomposition of carbon tetrafluoride.

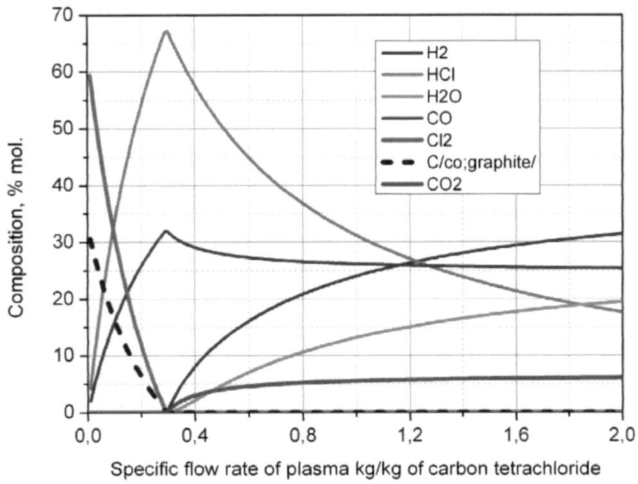

Figure 2. Decomposition of carbon tetrachloride.

3. Experimental part

Experimental study of carbon tetrachloride decomposition was conducted on the laboratory installation, which schematic is shown in figure 3 using three-phase AC plasma torch (figure 4). This is a high-voltage plasma torch with rod electrodes installed in the individual cylindrical channels. The plasma forming gas (or gas mixture) can be supplied into the near electrode region and into discharge channels.

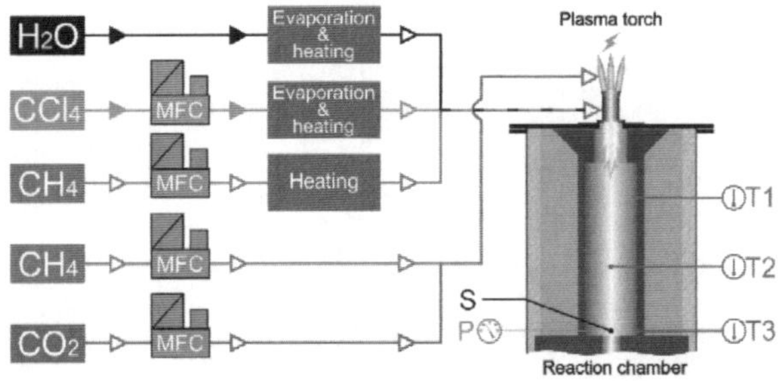

Figure 3. Schematic diagram of the experimental installation. T1, T2 and T3 – thermocouples, P – pressure gauge, S – sampling point.

Figure 4. Three-phase AC plasma torch.

More detailed description of the plasma torch, power supply and systems of electrical parameters measurement is presented in [16]. Liquid carbon tetrachloride is fed into the evaporator-superheater where it is evaporated and heated to temperature of about 150 °C, after which its vapors are mixed with the preheated methane and steam. The resulting mixture is fed into discharge channels of the plasma torch. At the same time a part of the reaction components is fed as a protective gas (carbon dioxide and small amount of methane) into the near electrode region. This allows adjusting the plasma torch power by changing the arc voltage drop at constant electric current. Thermal plasma obtained from these mixtures is sent into the lined reactor where the final conversion of carbon tetrachloride takes place.

Selected product gas is cooled, and condensate is removed. The condensate is analyzed for the hydrogen chloride content, and the cooled product gas is analyzed for basic composition by mass spectrometer (Cirrus 300) and FTIR spectrometer (Nicolet 380).

4. Results and discussion

Experiment on destruction of carbon tetrachloride started from feeding of 0.8 g/s of methane into the near electrode region, and 0.41 g/s of methane into discharge channels of the plasma torch. The total methane flow rate was 1.21 g/s. Plasma torch power was increased to ~ 100 kW. Then 2.44 g/s of CCl_4 vapors were fed. Model agent was supplied for about 15 minutes. Flow rates, rms voltage and power are presented in table 1. Composition dried reaction products from the sampling point S1 is presented in table 2.

Table 1. Parameters of plasma torch operation.

Parameter	CO_2+H_2O	$CO_2+H_2O+CH_4$	$CO_2+H_2O+CH_4+CCl_4$
Flow rate of CO_2/H_2O, g/s	2.9/2.9	2.9/2.9	2.9/2.9
Flow rate of CH_4, g/s	0	1.21	1.21
Flow rate of CCl_4 g/s	0	0	2.44
rms voltage, V	1000	1215	1390
Electric current, A	52.0	52.2	51.2
Power, kW	84.1	101.5	116.7

Table 2. Product gas composition.

CH_4	H_2	CO_2	CO	HCl
1.84% vol.	32.46% vol.	33.62% vol.	12.97% vol.	19.13% vol.

The average power of the plasma torch was ~117 kW and temperature in the reaction volume was ~ 1300-1400 °C. Thus, the specific energy consumption per 1 kg of decomposed CCl_4 in this experiment was ~13.3 kWh/kg. Available means of measurement did not detect the initial substance in the reaction product composition, therefore it can be assumed that the conversion was close to 100%. Furthermore, hydrogen chloride was the only chlorine-containing substance in the reaction products.

As far as the model substance flow rate in the experiment was less than stoichiometric, reaction products have significant amount of initial substances (CO_2, CH_4 and H_2O). It is impossible to avoid the presence of these components (except for methane) in the real process products, because such processes should be carried out with excess of an oxidant (steam).

5. Conclusion

On the whole the performed experiment confirmed the possibility of using a high-voltage three-phase AC plasma torch for processing of organochlorine compounds. Experimental results show that hydrogen chloride is the only chlorine-containing compound during carbon tetrachloride processing. Total content of H_2 + CO in product gas is about 45.5% vol.

Acknowledgements
The work is partially supported by the RFBR grant 15-08-05909-a.

References

[1] Laube J, Keil A, Boenisch H, Engel A, Roeckmann T, Volk C and Sturges W 2013 Atmos. Chem. Phys. **13** 2779–2791
[2] Brown A, Volk C, Schoeberl M, Boone C and Bernath P 2013 Atmos. Chem. Phys. **13** 6921–6950
[3] Weber L, Boll M and Stampf A 2003 Crit. Rev. Toxicol. **33** 105-136
[4] Leylegian J, Law C and Wang H 1998 Twenty-Seventh Symposium (International) on Combustion (Pittsburgh) 529–536
[5] Hawari J, Davis S, Engel P, Gilbert B and Griller D 1985 J. Am. Chem. Soc. **107** 4721–4724
[6] Shuprov S and Tenser P 1999 Combust. Explos. Shock Waves **35** 386–392
[7] Bae J, Lee J and Lee K 2007 Ind. Eng. Chem. Res. **46** 7057–7065
[8] Aghsaee M, Darkon A, Eremin A, Duerrstein S, Boehm H, Somnitz H, Fikri H and Schulz C 2013 PCCP **15** 2821–2828
[9] Indarto A, Choi J, Lee H and Song H 2006 J. Environ.Sci. **18** 83–89
[10] Penetrante B, Hsiao M, Bardsley J, Merritt B, Vogtlin G, Kuthi A, Burkhart C and Bayless J 1997 Plasma Sources Sci. Technol. **6** 251–259
[11] Jędrzejczyk T, Kołaciński Z, Koza D, Raniszewsk G, Szymański Ł and Wiak S 2015 OpenChem. **13** 156–160
[12] Rubio S, Quintero M and Rodero A 2011 J. Hazard. Mater. **186** 820–826
[13] Krawczyk K and Ulejczyk B 2003 Plasma Chemistry and Plasma Processing **23(2)** 265-281
[14] Gaikwad V, Kennedy E, Mackie J, Holdsworth C, Molloy S, Kundu S, Stockenhuber M and Dlugogorski B 2014 Journal of Hazardous Materials. **280** 38-45
[15] Rutberg P, Kuznetsov V, Popov V, Popov S, Surov A, Subbotin D and Bratsev A 2015 Applied Energy **148** 159–168
[16] Rutberg P, Nakonechny G, Pavlov A, Popov S, Serba E and Surov A 2015 Journal of Physics D:Applied Physics **48** 245204-245215

High voltage AC plasma torches with long electric arcs for plasma-chemical applications

A V Surov[1, 2], S D Popov[1, 2], E O Serba[1], A V Pavlov[1], Gh V Nakonechny[1], V A Spodobin[1], A V Nikonov[1], D I Subbotin[1] and A M Borovskoy[1]

[1] Institute for Electrophysics and Electric Power of Russian Academy of Sciences (IEE RAS), Dvortsovaya emb. 18, 191186, St. Petersburg, Russia
[2] Peter the Great St.Petersburg Polytechnical University, Polytechnicheskaya 29,195251, St.Petersburg, Russia

E-mail: alex_surov@mail.ru

Abstract. Powerful AC plasma torches are in demand for a number of advanced plasma chemical applications, they can provide high enthalpy of the working gas. IEE RAS specialists have developed a number of models of stationary thermal plasma torches for continuous operation on air with the power from 5 to 500 kW, and on mixture of H_2O, CO_2 and CH_4 up to 150 kW. AC plasma torches were tested on the pilot plasmachemical installations. Powerful AC plasma torch with hollow electrodes and the gas vortex stabilization of arc in cylindrical channels and its operation characteristics are presented. Lifetime of its continuous operation on air is 2000 hours and thermal efficiency is about 92%, the electric arc length between two electrodes of the plasma torch exceeds 2 m.

1. Introduction

1.1. Plasma torches for plasma chemical applications

Apart from catalytic processes, non-catalytic high-temperature processes can be widely used in the future. Their main features are high rate of chemical reactions, conversion of feedstock and product selectivity [1]. For this effect, the powerful plasma torches may be used to produce thermal plasma, which is the energy source for chemical transformation. The main fields of application of high-power plasma torches in the chemical industry are solid waste gasification [2]; reforming of natural gas [3] and biogas [4]; radioactive waste processing [5]; plasma assisted combustion of fuels [6]. Thus, the thermal plasma can be regarded as a heat carrier medium [7] or chemical process feedstock (air [8], steam [9], CO_2 [10]). In any case, all these processes require high power, long continuous operation lifetime and plasma parameters constancy in industrial applications [11]. These problems limit an implementation of plasma-chemical technologies.

Effective thermal plasma sources are electric arc plasma torches (thermal plasma generators). The most widely used are direct current (DC) devices; overview of the various modern models is presented in [12]. Plasma torches operating in the power up to 20 MW have been designed. Continuous operation of plasma torches is limited by resource of electrodes, with the highest erosion occurring on the cathodes. The thermal efficiency of DC plasma torches usually does not exceed 80%, and the ongoing task of improving their effectiveness does not lose its relevance. Overview of alternative current (AC) systems designed for industrial applications is presented in [13]. Power range is up to 3 MW. Working gases used for DC and AC plasma torches are Ar, He, N_2, H_2, Air, CO_2, CO, CH_4 and steam. Using AC allows reduction of specific electrode erosion, as each electrode is alternatively an anode or cathode and to achieve high performance resource factors and high thermal efficiency.

1.2. High voltage AC plasma torches and plasma chemical installations developed in IEE RAS

Single-phase and three-phase high-voltage AC air plasma torches (50 Hz) were developed and investigated in the Institute for Electrophysics and Electric Power of Russian Academy of Sciences (IEE RAS). These are devices with long electric arcs, burning in separate cylindrical channels with rod electrodes [14]. Power of three-phase plasma torches is up to 75 kW, channel length for different models is ~ 0.2 - 0.5 m. Life time of water-cooled rod electrodes of these devices is up to 200 hours. The operating current is up to 30 A, voltage drop is up to 1700 V, air flow rate is up to 35 g/s, thermal efficiency is 95% and average temperature is 1500 - 5500 °C. Single-phase plasma torches operate in the power range up to 15 kW with air flow rate up to 4 g/s [15]. The operating current of such devices is up to 10 A, voltage drop is up to 2000 V, thermal efficiency is ~ 95%, and life time is 300 hours.

A high-voltage multigas arc three-phase AC plasma torch (50 Hz) with power up to 150 kW, operating on such plasma-forming media as steam, air, carbon dioxide, methane, vapors of halogenated organic compounds, as well as various mixtures of these gases and vapors have been developed in IEE RAS based on the experience in the development of air plasma torches with rod electrodes. It also used the scheme with long arcs (up to 0.8 m), burning in the separate discharge channels with rod water-cooled electrodes. Operating parameters of multigas plasma torch working on mixture of steam and air in different ratios have been studied [16, 17]. Dependences of its power characteristics, thermal efficiency on the arc current, flow rate and composition of the plasma-forming mixture were determined. Steam flow rate is 3.55 - 5.8 g/s, air flow rate is 1 - 3 g/s, arc current is 28 - 29 A, arc voltage drop is 1.15 - 1.85 kV, power is 57.6 - 87.5 kW, thermal efficiency is 94.3 - 95.3%. Also the parameters of the electric arc column (average for the arcing period) are determined, such as arc length (~ 0.8 m), bright arc core diameter (~4.47 mm), arc column temperature (10.0 - 11.5) $\times 10^3$ K, electric field strength (14 - 23 V/cm). Paper [18] presents the investigation results of the plasma torch operation on H2O/CO2/CH4 mixture under constant flow rates of steam (3 g/s) and carbon dioxide (3 g/s) and with methane flow rate varying from 0 to 0.8 g/s. Operating characteristics depending on arc current, flow rate and composition of plasma forming mixture were obtained. Arc current is 52 – 59 A, arc voltage drop is 0.87 – 1.13 kV, power is 77.7 – 114.6 kW, thermal efficiency is 93.8 – 95.9%, average temperature of plasma is (3.2 – 3.3) $\times 10^3$ K. Life time of continuous operation of electrodes is 200 hours. Also the parameters of the electric arc column (average for the arcing period) are determined, such as the length (~ 0.6 m), bright arc core diameter (5.99 – 7.54 mm), arc column temperature (8.3 – 10.5) $\times 10^3$ K, electric field strength (14 – 22 V/cm).

The developed AC plasma torches were tested on IEE RAS pilot plasma chemical installations for gasification of solid waste and reforming of gaseous hydrocarbons. Air plasma torches with rod electrodes worked in the downdraft gasifier during investigation of air gasification of wood waste. Thus, due to the heat of plasma the syngas quality was considerably improved in comparison with autothermal gasification: content of combustible components (H_2 + CO) in the product gas was 55 – 67%, nitrogen was 26 – 36%; LHV (lower heating value) of syngas was 5.9 – 7.2 MJ/Nm^3; content of tars was 50 mg/m^3. Air plasma gasification of organic solid matter flows at relatively low specific energy consumptions (2.4 – 3.2 MJ/kg). The experimental results are represented in [19].

Similarly, the multigas plasma torch worked with the plasma-chemical reactor for the methane reforming by steam and carbon dioxide [20]. Due to flexible regulation of the plasma-forming medium composition without shutting down the torch and change the power supply settings it was possible to obtain synthesis gas with H_2/CO ratio of 1 to 2.4. Hydrogen and carbon monoxide selectivities were 98%, the methane conversion was 99%, acetylene and soot content did not exceed 0.1 %vol. and

125 mg/Nm^3 of synthesis gas respectively. Specific power consumption of the process depends on the plasma forming gas composition: for steam/carbon dioxide reforming is 37.5 MJ/kg of converted methane; for dry reforming is 48.3 MJ/kg of converted methane.

High temperature of the electric arc does not only have a positive effect on working gas heating, but also introduces a number of limitations on usage of some design concepts. Rod electrodes have good resource characteristics when operating on air with currents do not exceeding 30 A. When operating from the AC power supply (50 Hz) with no-load voltage of 10 kV it was possible to obtain a steady burning arc with voltage drop of 3 kV. Investigations on development of electrodes with a long life time at high currents (up to 100 A) were carried out in order to create a plasma torch with power more than 150 kW. Such characteristics when operating on alternating current were obtained on cylindrical electrodes with internal working surface (hollow electrodes). The operation of experimental models of hollow electrodes was investigated to determine the possibility of increasing the resource. The modes allowing distribute thermal impact at controlled movement of the electric arc spot along electrodes surface under the influence of gas dynamic forces and magnetic field were determined. [21, 22].

Investigations on the experimental installation equipped with supply systems, mass flow controllers for plasma forming gases, acquisition of electrical parameters, high-speed video and spectral diagnostic equipment were carried out to create a powerful plasma torch. A series of experiments for different power levels and air flow rates were performed. The powerful AC plasma torch with hollow electrodes was developed on the basis of obtained experimental data. Its characteristics will be presented below.

2. Experimental setup for investigations of AC electric arc systems with hollow electrodes

The operation of single- and three-phase experimental models of plasma torch with hollow electrodes was investigated to obtain characteristics for the high power plasma torch development. Electric field strength of the arc column at low current (up to 10 A) in 10 kW air plasma torch is about 30-40 V cm^{-1}, but one is only 7-11 V cm^{-1} for electric arcs with current of about 80 A. When operating at current of about 80 A to produce a high arc voltage drop it was required to increase the length of the channels (up to ~ 1 m). Photo of running experimental single-phase model of the plasma torch with the long discharge channels and various designs of hollow electrode units is presented in figure 1.

(a)

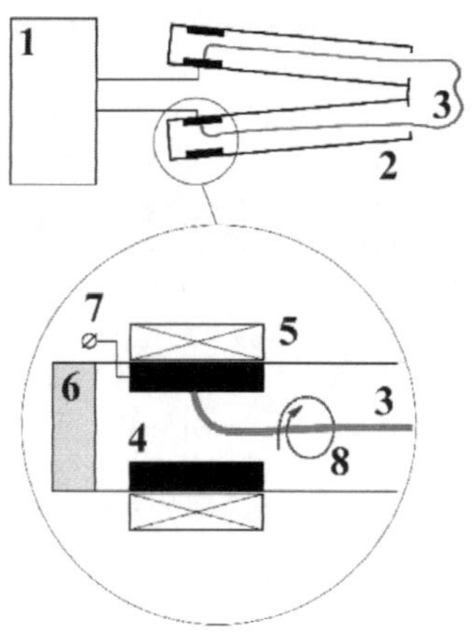

(b)

Figure 1. Experimental single-phase model of high-voltage AC plasma torch. (a) Photo of operation, (b) schematic diagram of device. 1 – Power supply; 2 – case; 3 – arc; 4 – electrode; 5 – solenoid; 6 – diagnostic window; 7 – terminal for power supply connection; 8 – tangential working gas flow.

IOP Conf. Series: Journal of Physics: Conf. Series **825** (2017) 012016 doi:10.1088/1742-6596/825/1/012016

The single-phase electric arc has spots on two electrodes disposed in separate channels. The position of the arc spot and arc column around the electrodes is controlled by the magnetic field and the swirling gas flow. Most part of the arc column length is located along the axes of the discharge channels and its small section in the external space. So the electric arc length between two electrodes of experimental model is about 2,5 m (1 m + 1 m in channels + ~ 0.5 m outside).

The measuring installation including electrical probes, acquisition system for AC high voltage signals measurement with a sampling rate of no less than 10 kHz was created to determine the strength of the electric field in different areas of the arc column and plasma parameters. Testing of elements of the measuring system with video recording was carried out.

Burning of arcs in channels at three-phase operation mode is the same as at single-phase, the arc columns at the outlets of three channels are closed to each other in the surrounding area. High-speed video and photo shooting were used to investigate the character of electric arcs burning at the outlet of the channels. The schematic diagram of three-phase AC plasma torch and arrangement of photo and video cameras are presented at figure 2 (a). The arcs at the channel outlets are represented at figure 2 (b) – photo from side camera (9-S at schematic diagram), figure 3 (a) shows the arc in the near electrode area (rear view camera 9-R).

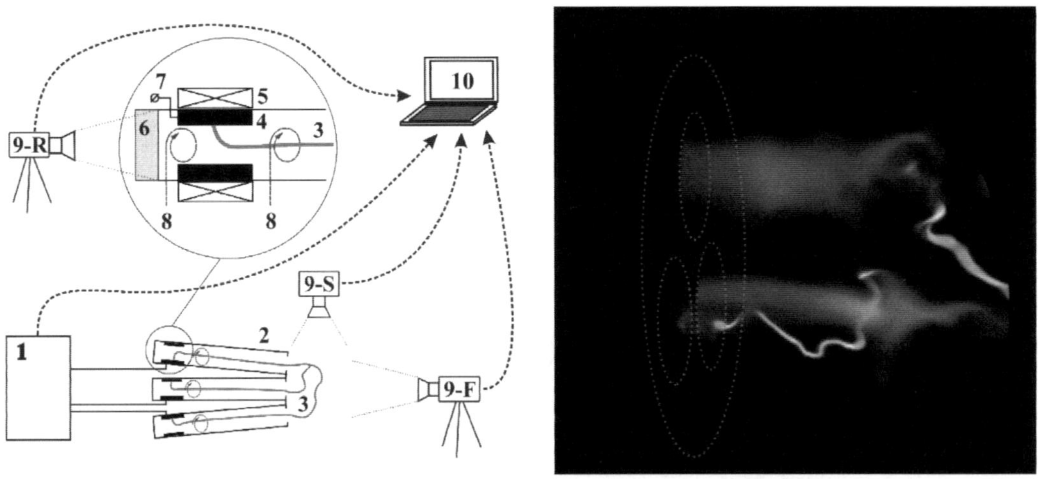

(a) (b)

Figure 2. (a) Schematic diagram of high voltage three-phase AC plasma torch with hollow electrodes. (b) Photo of electric arcs outside the channels – side view, contour line indicates boarders of front part of the plasma torch and outputs of channels (shutter speed 1/8000 s). 1 – Power supply; 2 – case; 3 – arc; 4 – electrode; 5 – solenoid; 6 – diagnostic window; 7 – terminal for power supply connection; 8 – gas supply; 9-R, 9-S and 9-F – photo/video camera position (rear, side and front view accordingly); 10 – data acquisition system.

Arc column has a relatively small diameter of the light emitting area in the cold air flow at the electrode surface, the arc within the channels burns in the axial area and has a larger diameter. It is evident that the arc just in front of the channels outlet also has a relatively large diameter (in the picture looks fuzzy), but at the longer distance from the case, the part of the arc column captured by the transverse gas flow is contracted. Figure 3 (b) shows a photo of the electrode working surface.

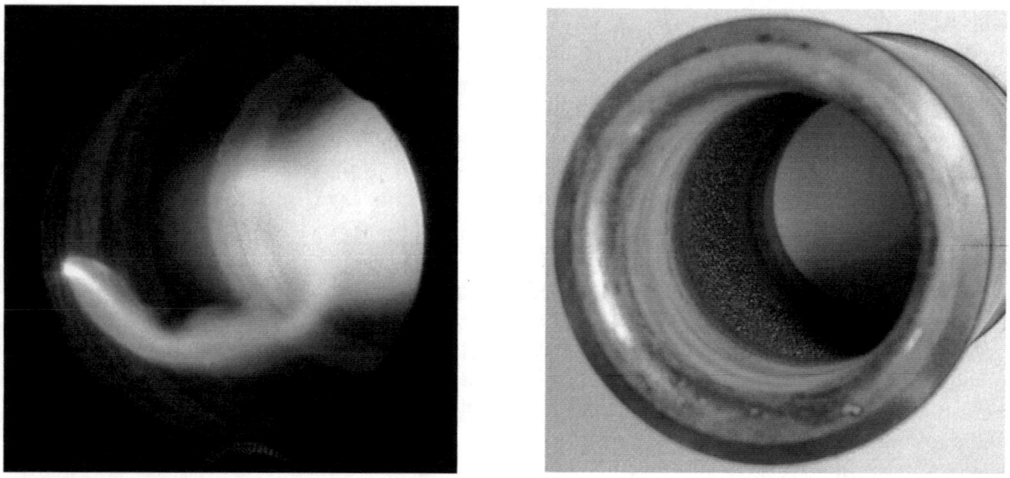

(a) (b)

Figure 3. (a) Photo of electric arc in AC plasma torch with hollow electrodes: arc spot on the electrode surface, shutter speed 1/4000 s [19]; (b) photo of the electrode working surface after 180 hours of operation with current 83 A.

3. High power three-phase plasma torch with hollow electrodes

The powerful three-phase AC air plasma torch with hollow electrodes operating in the power from 150 to 500 kW was developed on the basis of the performed experiments. Photo of the plasma torch is shown in figure 4.

Figure 4. High-voltage AC plasma torch. Operation power 390 kW.

This is a device with gas vortex stabilization of the arc in cylindrical channels. The electric arc length between two electrodes of the plasma torch is about 2.5 m (1 m +1 m in channels + ~ 0.5 m outside, as shown at figure 2 (a)). It allows working with a high arc voltage drop (~ 2 – 3 kV) at relatively low currents (no more than ~100 A) what favorably affects the lifetime of electrodes. The operation principle of the plasma torch is the following: arcs stabilized by the swirling gas flow burn in separate cylindrical channels, close on each other at the outlet of the case. The arc spots are continuously rotated along the inner surface of the hollow electrode. Continuous scanning of the arc spots along the electrode surface in up-streem/down-streem directions is carried out to increase the life time of electrodes (simultaneously with the arc spot rotation). Preliminary short-term tests of plasma torch operation (~ 5 minutes) with power up to 500 kW (103 A, 2900 V) were performed. Characteristics of plasma torch were obtained during long-term tests (more than 30 minutes for each mode) for the arc current up to ~ 85A (up to ~410 kW). The thermal efficiency of the plasma torch is ~ 92%.

Life time tests of the plasma generator were carried out, performance characteristics were obtained. Figure 5 shows the static volt-ampere characteristic (VAC) of the arc, burning in the plasma torch at air flow rate of 65 g/s.

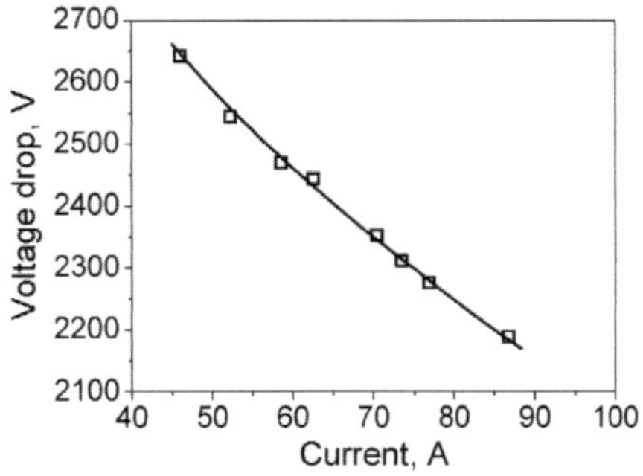

Figure 5. Static VAC of the arc of high power AC plasma torch with hollow electrodes.

The curve has a falling character, which is due to dropping VAC of the power supply system. The power of the plasma torch varies from ~ 200 to 330 kW at concerned current range. Possibility of significant changes in power at fixed flow rate of the plasma-forming medium and at the expense of the current change allows control of the specific energy consumption over a wide range.

Figure 6 represents the dependences of arc voltage drop and power versus air flow rate for arc current of ~ 85 A. The RMS values of arc current and the voltage drop on the arc and the active power of plasma torches were automatically calculated during operation on the instantaneous values of currents and voltages. Apparently, the arc voltage drop and power with increasing of air flow rate significantly increase. This is due to increasing of the intensity of heat exchange of the gas flow with the electric arc. Increasing in the heat exchange intensity causes cooling and contraction of the arc column, leading to the increase in current density and reduction in electroconductivity.

Figure 7 shows the dependences of enthalpy and average temperature of plasma versus air flow rate for the same operational modes of the plasma torch.

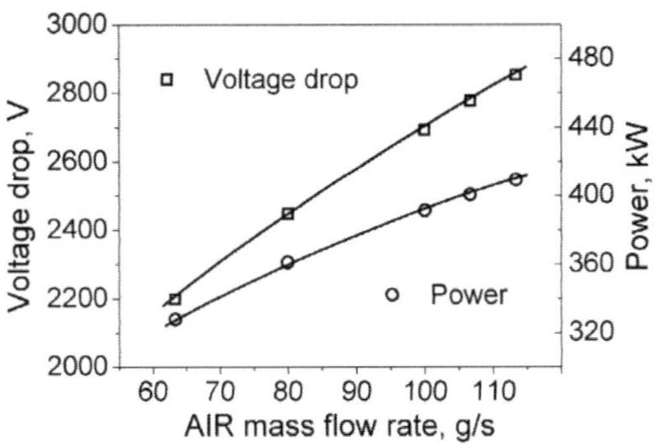

Figure 6. Power and arc voltage drop of high power AC plasma torch with hollow electrodes.

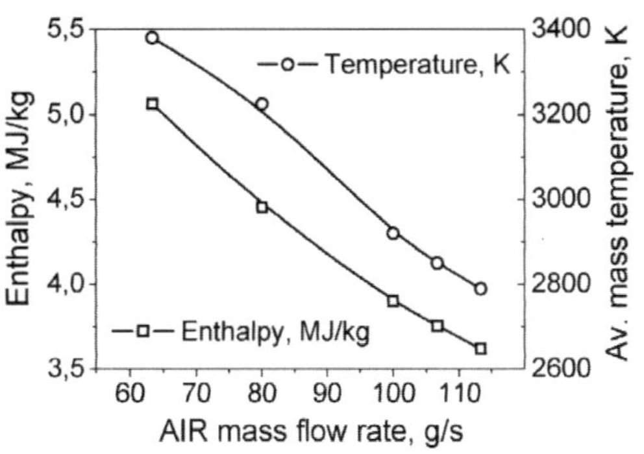

Figure 7. Enthalpy and average temperature of plasma of high power AC plasma torch with hollow electrodes.

These charts illustrate the flexibility and possibility of operational management of plasma flow rate and its enthalpy in a wide range. That may be useful for controlling of the plasma-chemical processes parameters.

Figure 8 shows the presented frames of high-speed shooting of high-power plasma torch operation (front view). Diffuse and contracted areas of the arcs at the channel outlets are visible. More bright arcs can be seen in the moments of current peaks, less bright arcs are observed at transition of the corresponding phase current through zero.

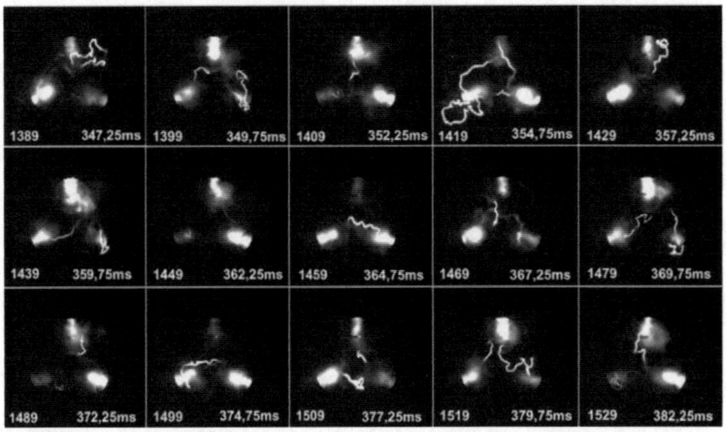

Figure 8. Arcs of AC plasma torch with hollow electrodes. Frames of high speed video (frame rate 4000 fps, the exposure time of each frame is 2 μs). Each frame has its number on the left, time from start of record is on the right.

The life time of electrodes determined as a result of certain tests at current of 83 A was about 2000 hours. Erosion of electrodes on the results of weighing after the first 50 hours of operation was 1.1×10^{-6} g/C, then its value gradually increased and stabilized at a value of 3.3×10^{-6} g/C at 180 hours of operating time of the electrodes. Then the erosion did not increase. The pattern of wear of the electrodes with current 83 A is shown above in figure 3 (b).

4. Conclusions

AC plasma torches with long electric arcs have good performance characteristics and are promising for the plasma-chemical applications. The usage of these devices as part of the pilot plasma chemical plants ensured the success of the experiments on the air plasma gasification of organic substances. For wood waste the content of fuel gas (H_2 + CO) in the products was 55-57%.

The AC plasma torch with power up to 500 kW has been developed. Its operation with arc currents up to ~85 A and power up to 410 kW was investigated. Organization of the long arc burning in the plasma torch channels with voltage drop of 2800 V allows operate with a relatively low current and obtain service life of ~2000 hours on air.

Acknowledgements
The work is supported by the RFBR grant 15-08-05909-a.

References

[1] Fridman A 2008 *Plasma Chemistry* (Cambridge: Cambridge University Press)
[2] Evangelisti S, Tagliaferri C, Clift R, Lettieri P, Taylor R and Chapman C 2015 Integrated gasification and plasma cleaning for waste treatment: A life cycle perspective *Waste Management* **43** 485–96

[3] Rutberg Ph G, Kuznetsov V A, Popov V E, Popov S D, Surov A V, Subbotin D I and Bratsev A N 2015 Conversion of methane by CO_2+H_2O+CH_4 plasma *Applied Energy* **148** 159–68
[4] Rueangjitt N, Akarawitoo C and Chavadej S 2012 Production of Hydrogen-Rich Syngas from Biogas Reforming with Partial Oxidation Using a Multi-Stage AC Gliding Arc System *Plasma Chem. Plasma Proc.* **32** 583–96
[5] Komatsu F, Takusagawa A, Wada R and Asahina K 1990 Application of microwave treatment technology for radioactive wastes *Waste Management* **10** 211–15
[6] Ju Y and Sun W 2015 Plasma assisted combustion: Dynamics and chemistry *Progress in Energy and Combustion Science* **48** 21–83
[7] Tao X, Qi F, Yin Y and Dai X 2008 CO_2 reforming of CH_4 by combination of thermal plasma and catalyst *Int. J. Hydrogen Energy* **33** 1262–65
[8] Bromberg L, Cohn D R, Rabinovich A and Alexeev N 1999 Plasma catalytic reforming of methane *Int. J. Hydrogen Energy* **24** 1131–37
[9] Tamošiūnas A, Valatkevičius P, Grigaitienė V, Valinčius V and Striūgas N 2016 A cleaner production of synthesis gas from glycerol using thermal water steam plasma *Journal of Cleaner Production* **130** 187–94
[10] Bo Z, Yan J, Li X, Chi Y and Cen K 2008 Plasma assisted dry methane reforming using gliding arc gas discharge: Effect of feed gases proportion *Int. J. Hydrogen Energy* **33** 5545–53
[11] Harry J E 2010 *Introduction to Plasma Technology: Science, Engineering and Applications* (Weinheim: Wiley-VCH Verlag GmbH & Co. KGaA)
[12] Mostaghimi J and Boulos M I 2015 Thermal Plasma Sources: How Well are They Adopted to Process Needs? *Plasma Chem. Plasma Proc.* **35** 421–36
[13] Fulcheri L, Fabry F, Takali S and Rohani V 2015 Three-Phase AC Arc Plasma Systems: A Review *Plasma Chem. Plasma Proc.* **35** 565–85
[14] Rutberg Ph G, Safronov A A, Popov S D, Surov A V and Nakonechny Gh V Multiphase stationary plasma generators working on oxidizing media *Plasma Phys. Control. Fusion* **47** 1681–96
[15] Rutberg F G, Pavlov A V, Popov S D, Sakov A I, Serba E O, Spodobin V A and Surov A V 2009 Spectral Measurements of the Gas and Electron Temperatures in the Flame of a Single-Phase AC Plasma Generator *High Temp.* **47** 175–180
[16] Rutberg Ph G, Kuznetsov V A, Serba E O, Popov S D, Surov A V, Nakonechny Gh V and Nikonov A V 2013 Novel three-phase steam-air plasma torch for gasification of high-caloric waste *Applied Energy* **108** 505–14
[17] Rutberg Ph G, Kuznetsov V A, Serba E O, Nakonechnyi G V, Nikonov A V, Popov S D and Surov A V 2013 Study of electric arcs in an air–steam mixture in AC plasma torches *High Temp.* **51** 608–14.
[18] Rutberg Ph G, Nakonechny Gh V, Pavlov A V, Popov S D, Serba E O and Surov A V 2015 AC plasma torch with $H_2O/CO_2/CH_4$ mix as the working gas for methane reforming *J. Phys. D: Appl. Phys.* **48** 245204
[19] Rutberg Ph G, Bratsev A N, Kuznetsov V A, Popov V E, Ufimtsev A A, Shtengel' S V 2011 On efficiency of plasma gasification of wood residues *Biomass and Bioenergy* **35** 495–504
[20] Rutberg Ph G, Bratsev A N, Kuznetsov V A, Nakonechny G V, Nikonov A V, Popov V E, Popov S D, Serba E O, Subbotin D I and Surov A V 2014 Production of synthesis gas by conversion of methane in a steam-carbon dioxide plasma Tech. Phys. Lett. **40** 725–9
[21] Rutberg Ph G, Popov S D, Surov A V, Serba E O, Nakonechny Gh V, Spodobin V A, Pavlov A V and Surov A V 2012 The investigation of an electric arc in the long cylindrical channel of the powerful high-voltage AC plasma torch *J. Phys.: Conf. Ser.* **406** 012028
[22] Surov A V, Popov S D, Serba E O, Nakonechny Gh V, Spodobin V A, Ovchinnikov R V, Kumkova I I and Shabalin S A 2012 The investigation of movement dynamics of an AC electric arc attachment along the working surface of a hollow cylindrical electrode under the action of gas-dynamic and electromagnetic forces *J. Phys.: Conf. Ser.* **406** 012007

Suitability of thermal plasmas for large-area bacteria inactivation on temperature-sensitive surfaces – first results with Geobacillus stearothermophilus spores

M Szulc[1]*, S Schein[2], J Schaup[2], S Zimmermann[2] and J Schein[2]

[1] Zierhut Messtechnik GmbH, Munich, Germany
[2] Lab for Plasma Technology (LPT), Universität der Bundeswehr München, Neubiberg, Germany

* szulc@zierhut-messtechnik.de

Abstract. The application of thermal plasma for large-area bacteria inactivation on temperature-sensitive surfaces is not a common one. Nonetheless, there are thermal plasma generators which offer a high sheath homogeneity and have proven to be suitable for treatment of thermally sensitive materials in the past. To investigate the suitability of such plasmas, agar dishes plated with endospores of Geobacillus stearothermophilus have been treated with a long arc plasma generator called LARGE. The achieved results have been compared with a commercially available non-thermal plasma generator. A significant inactivation of the endospores could be observed only after 60 s of treatment with the thermal plasma source. This was not possible with the non-thermal generator. Moreover, no temperature damage or increase of the specimen could be detected. An attempt to determine the main agents responsible for the microbicidal effects have been made – the influence of plasma gas composition, discharge current and treatment time has been investigated. Significant improvements in the disinfection rates after adding small amounts of nitrogen to the plasma gas could be observed. A first discussion regarding the suitability of thermal plasmas for bacteria inactivation has been given.

1. Introduction

Treatment with non-thermal plasmas is by now a well-established method for bacteria inactivation in medical and biological research and has been investigated by many researchers [1, 2, 3]. As stated in the literature various plasma agents and properties are responsible for the sterilisation effect. In general, the main agents are: temperature, UV radiation, reactive nitrogen and oxygen species (short RNS and ROS, respectively) and short-lived charged particles [4, 5]. Although the sterilizing agents can be defined, many researchers state that the interaction effects between plasma and bacteria are not fully understood [2, 3, 6, 7]. This may be due to insufficient or missing plasma diagnostics. A major challenge thereby is to quantify the reactive and therefore very short-lived species, whose concentration is often very low and decays rapidly in effluent plasma jet under atmospheric pressure [8, 9].

The application field of such plasmas can be manifold. They are often used as an alternative to traditional sterilization methods of stents, implants, scalpels or other surgical instruments [7, 10], for bacteria inactivation [11, 12] or wound healing [13, 14]. Moreover, investigations in the fields of water and air decontamination [15, 16], treatment of heat- and humidity-sensitive food as well as inactivation of chemical and biological warfare agents [17, 18] have been performed by various researcher groups. There are even some commercially available CE-certified atmospheric pressure cold plasma sources for medical treatments [19, 9].

Most of the investigated non-thermal plasmas have a jet- or pen-shaped geometry, which allows a point-like surface treatment with a diameter of several mm^2. Nonetheless, concerning the application fields, a need for large-area surface decontamination can be easily recognized. Efforts to create a large-area surface sterilisation by upscaling a non-thermal discharge or stringing together several cold plasma jets had already been undertaken by various researchers [20, 21, 22]. The stability and plasma sheath homogeneity is one of the main concerns of such systems [1, 21].

Attempts to realize atmospheric pressure large-area plasma generators with high plasma sheath homogeneity have been successfully undertaken by some groups [23, 24]. These designs were based on thermal plasmas rather than non-thermal ones. Farrar et al. studied even the applicability of their thermal plasma generator for large-area surface sterilisation [23]. Investigated application were the treatment of runways, roads or buildings, all of those non temperature-sensitive surfaces. No further literature which described the application of thermal plasmas for bacteria inactivation on temperature-sensitive surfaces could be found.

One thermal plasma generator with high sheath homogeneity designed for large-area treatments, which has also shown good suitability for large-scale surface treatment of temperature-sensitive substrates in previous works [24, 25, 26], has been developed at the LPT and is called LARGE. Within this work, the suitability of thermal plasmas (produced by LARGE) for large-area bacteria inactivation on temperature-sensitive surfaces has been investigated.

2. Approach

Within this section the function principle of the used plasma sources has been described. Additionally, the choice of bacterium has been explained as it is of great importance for following results.

2.1. LARGE – Long ARc plasma GEnerator

LARGE, a long arc plasma generator, is a linear DC-plasma source with an up to 450 mm long electrical arc discharge, where the plasma gas is fed perpendicularly to the arc. A high temporal stability of its operating parameters, and hence of the plasma sheath, can be achieved, since the arc length is defined by the distance between the electrodes and the arc is being stabilized by water-cooled cascades and a magnetic barrier stretching along the whole arc length, as shown in figure 1. Insulating plates are placed between each element of the cascade to prevent a current flow through one of them, and thus force the arc to burn between the main electrodes. The cascaded layout facilitates also an easier arc ignition. The Lorenz force created by the two permanent magnets on the other hand, prevents an arc deflection caused by the lateral plasma carrier gas flow, figure 2. Such a gas injection allows the generation of thermal plasmas with different (also aggressive or oxidative) gases, such as Ar, N_2, O_2, compressed air, H_2, He, CH_4 or CO_2 [26].

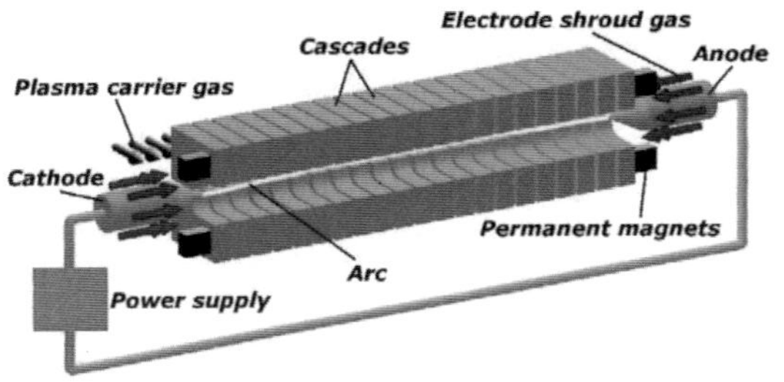

Figure 1. Schematic layout of the thermal plasma generator LARGE.

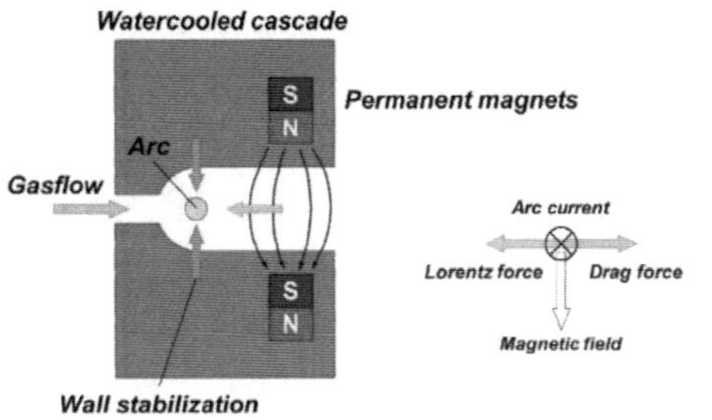

Figure 2. Cross-section of a LARGE cascade describing the arc stabilization principle

Depending on the gas type and chosen process parameters, temperatures between 100 °C and 4500 °C in the effluent plasma sheath at generator exit can be achieved. According to that, all main agents responsible for plasma enhanced decontamination can be generated and adjusted within a relatively wide range. Additionally, due to high arc stability, a large-area temporarily and spatially homogenous plasma sheath can be created. Because of its characteristics the LARGE seems to be suitable for large-area decontamination of temperature-sensitive surfaces and therefore should be used within this work.

2.2. Non-thermal plasma generator for comparison reasons - Relyon Plasma PB3

As a reference point the commercially available non-thermal plasma system, PB3 plasma generator and a PS2000 power supply, both produced by Relyon Plasma GmbH, Regensburg, Germany, has been used. According to the manufacturer this generator can be used for surface sterilization in industrial applications, for ex. packaging [27]. The plasma generator PB3 has a more conventional construction with a finger-shaped inner electrode and a nozzle acting as a cathode, as presented in figure 3. A unipolar pulsed high voltage (a few kV) arc discharge is used to generate a plasma jet. To prevent a pre-mature electrode wear, a vortex flow of the plasma carrier gas inside the nozzle is introduced and so the arc rotates inside the discharge chamber at a high frequency. Similarly to the LARGE, different working gases can be used, most commonly compressed air, N_2, CO_2 or N_2/H_2-mixtures. Furthermore, the plasma temperature can be adjusted by varying the nozzle length. The system does not require any additional cooling.

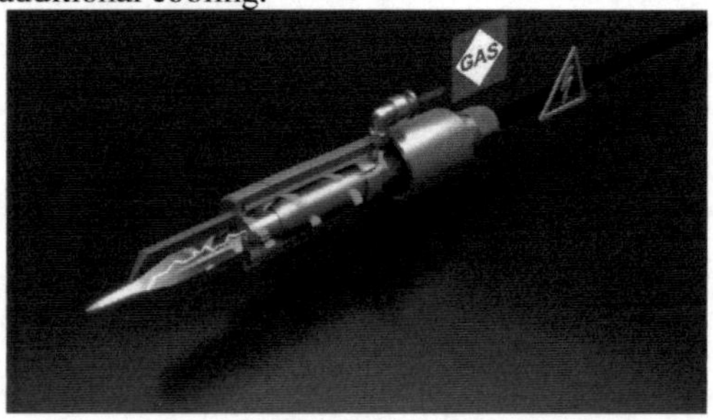

Figure 3. Schematic layout of the non-thermal plasma generator PB3 used for comparison reasons.

2.3. Bacteria choice - Geobacillus stearothermophilus ATCC 7953

The choice of bacterium used to determine the sterilisation effects of a plasma appears to be of particular importance. As shown by Morris et al. [28] significant differences in inactivation rates can be observed between a vegetative bacterium and its endospore. The authors examined the effects of treatment with a cold atmospheric pressure plasma (direct and indirect treatment) on Geobac. stearothermophilus and Bacillus cereus in both forms, endospore and vegetative. Geobac. stearothermophilus is a spore-forming, gram-positive rod-shaped aerobic bacterium, whose optimum growth temperature is between 50 °C and 65 °C [29]. During investigations the treatment time has been varied between 10 s and 30 min. No further information about the used plasma gas and other process parameters has been given. The results show that after treatment with the cold plasma both vegetative forms of Geobac. stearothermophilus and Bacillus cereus, as well as the Bacillus cereus spores regardless of whether plated on agar dishes or on glass specimen could be effectively inactivated. The endospores of Geobac. stearothermophilus however, could not be significantly inactivated by any sort of plasma treatment, neither indirect nor direct one.

According to Yardimci and Setlow [30] spores of Bacillus strain and their close relatives (Bac. atrophaeus, Bac. cereus, Bac. subtilis and Geobac. stearothermophilus) possess a significantly higher resistance against plasma treatment than the vegetative forms of these strains or other bacteria. This applies under the assumption of identical treatment conditions, even if different plasma types were used for the experiments. Also other comparative studies conclude that the endospores of Geobac. stearothermophilus are among the species with the greatest resistance to a plasma treatment and therefore can be used as a suitable biological indicator for the assessment of the microbicidal effect of a plasma [31].

Hence, spores of Geobac. stearothermophilus appear to be particularly well suited to prove the suitability of thermal plasmas for large-area bacteria inactivation on temperature-sensitive surfaces.

3. Experimental setup

To show the suitability of thermals plasmas for bacteria inactivation on temperature-sensitive surfaces a simple two-step analysis method (screening and quantification) has been applied. At first, tests with a dense bacteria layer have been conducted to screen the parameters and determine the main influencing factors. Therefor 50 µl of an undiluted base suspension containing 2.2×10^6 CFU/ml (CFU – colony forming unit) of Geobac. stearothermophilus ATCC 7953 spores have been plated onto tryptone soya agar (TSA) dishes and incubated for 24h at 56 °C after treatment. The main focus was to determine the influence of gas composition, treatment time and distance as well as input power, i.e. arc current or pulse frequency depending on the used plasma system.

After the determination of the influencing factors, disinfection rates have been determined by cell counting to quantify the plasma effects. To do so a 1:121 dilution of the base suspension (2.2×10^6 CFU/ml) has been used. Again, 50 µl of this dilution, which contain an average of 909 CFU, have been distributed onto TSA plates and incubated under similar conditions after the plasma treatment. The treatment distance for these investigations has been kept constant at 60 mm. Other parameters, set based on the results with dense bacteria layers, were examined and fine-adjusted if needed. Moreover, for statistical reasons, three samples have been analysed for each parameter set. The detailed LARGE parameters are presented in table 1.

Table 1. Parameters set for the investigations with thermal plasma generator LARGE.

	Process parameter:	Value:
constant	Arc length	150 mm
	Electrode shroud gas	Ar
	Shroud gas flow rate	2.5 l/min
	Total gas flow rate	35 l/min
	Treatment distance	60 mm
variable	Base plasma carrier gas type	Ar
	Additional plasma carrier gases	N_2, O_2
	Additional gas flow rate	5 l/min (14 Vol.-%), 10 l/min (28 Vol.-%)
	Arc current	40 A, 50 A
	Treatment time	15 s, 30 s, 45 s, 60 s, 120 s

The results achieved with the LARGE have been compared with a commercially available non-thermal plasma generator (Relyon Plasma PB3). All experiments with both plasma generators have been performed stationarily, i.e. without moving the dishes or the plasma generator, as shown in figure 4 and figure 5.

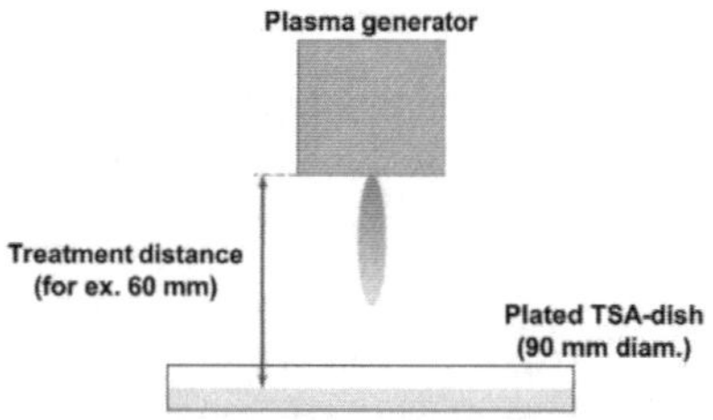

Figure 4. Schematic diagram of the testing conditions. Experiments have been performed stationary.

Figure 5. A plated TSA-dish is being treated with the thermal plasma generator LARGE. No thermal degradation could be observed.

4. Results

The results presented within the following section concentrate primarily on the results achieved with thermal plasmas, rather than non-thermal one.

4.1. Screening – determining the main influencing factors

First sterilization experiments with a dense microbial biofilm of Geobac. endospores showed that a significant amount of spores could be killed after just 60 s of treatment with the thermal plasma generator LARGE (figure 6). In comparison, a significant deactivation of the endospores could not be observed even after a four times longer treatment of 240 s with the non-thermal plasma generator PB3. Moreover, after such a long treatment with the cold plasma jet a thermal degradation of the agar dish could be observed. As shown in figure 7, a significant bacteria inactivation after a treatment with the PB3 generator could only be observed after a pre-incubation of plated dishes (6 h at 56 °C). Hence, only the vegetative form of Geobac. stearothermophilus could be effectively inactivated with the cold plasma, what coincides with the results obtained by Morris et al. [29]. For such testing conditions (vegetative form, static treatment), best results with the PB3 could be achieved in a treatment distance of 60 mm, as shorter distances caused a thermal damage of the substrates and longer ones lessened the effect. According to that, for comparison reasons, the distance of 60 mm has been kept unchanged for LARGE experiments.

The experiments showed also that plasma gas composition plays a key role in sterilization processes regardless of the used plasma generator type. In the case of the LARGE significant improvements can be achieved when small amounts of N_2 or O_2 are added to the plasma carrier gas (Ar). An arc current increase (from 40 A to 50 A) also improved the disinfection effects. The positive effect was however much lesser in comparison to gas composition change. Furthermore, compared to the non-thermal plasma jet, thermal degradation of the agar could not be observed even after 120 s treatments with the LARGE. The average gas temperature measured in the treatment distance of 60 mm was 23.6 °C (PB3 in comparison: 84.1°C). Furthermore, as one expected, better inactivation rates (to an extent) could be achieved by elongating the treatment time (see section 4.2).

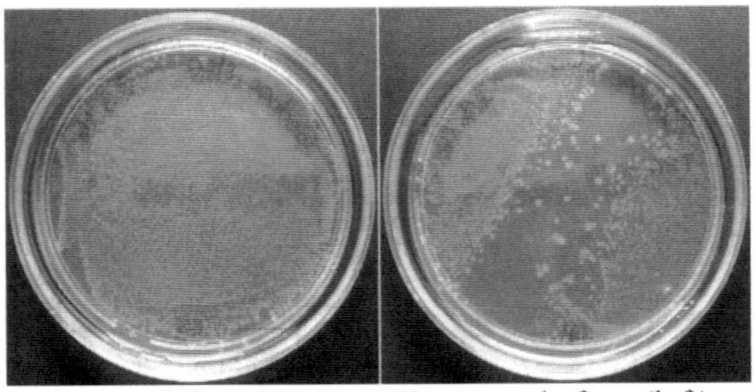

Figure 6. Geobac. stearothermophilus ATCC 7953 spores before (left) and after (right) 60 s of treatment with the LARGE. Parameters: plasma carrier gas flow: 20 l/min Ar + 10 l/min N_2, electrode shroud gas flow: 1.5 l/min, arc current: 50 A, treatment distance: 60 mm

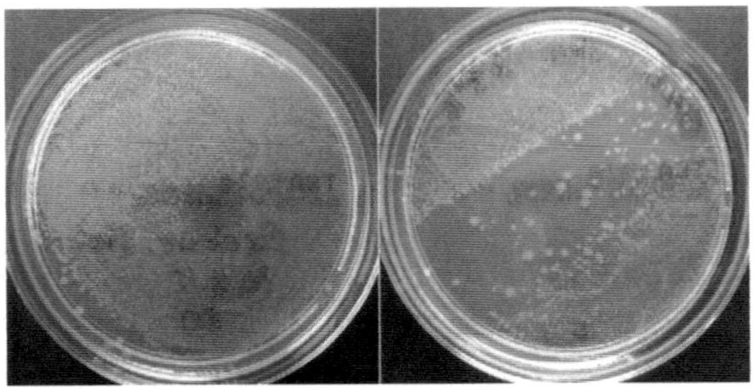

Figure 7. Comparison of results after 60 s of treatment. Left: non-thermal plasma (PB3), dish pre-incubated after plating for 6h at 56°C. Right: thermal plasma (LARGE), direct treatment of endospores

In summary it can be said, that the thermal plasma (produced by the LARGE) seems to be suitable for a large-area bacteria inactivation. Moreover, the key parameters in this application field are the plasma gas composition followed by the treatment time and the arc current.

4.2. Quantification – determining the disinfection rates

To quantify the antimicrobial impact of the thermal LARGE-plasma, a dilution of the base suspension containing an average of 909 CFU has been used. As previously conducted tests show, plasma carrier gas composition and treatment time have the highest impact on bacteria inactivation. Moreover a significant improvement could be observed when small amounts of nitrogen or oxygen were added. Accordingly to that, quantification tests with pure argon as plasma carrier gas have been skipped. A N_2-O_2 mixture has not been investigated as it would complicate the differentiation of the plasma agents responsible for the sterilization effect. The determined disinfection rates for different gas compositions have been shown in figure 9. Best results for a constant treatment time of 60 s could be achieved for nitrogen (5 l/min = 14 Vol.-%). Therefore, the treatment time study has been conducted for an Ar-N_2 (14 Vol.-%) plasma gas mixture and an arc current of 50 A. Other parameters have been set according to table 1. As shown in figure 8, a significant degradation of CFUs could be observed just after 30 s of treatment. Further extension of the plasma exposition time increases the impact only marginally. Noteworthy, as stated already before, no thermal degradation could be observed during these tests. In both figures the untreated samples have been marked as reference (Ref).

IOP Conf. Series: Journal of Physics: Conf. Series **825** (2017) 012017 doi:10.1088/1742-6596/825/1/012017

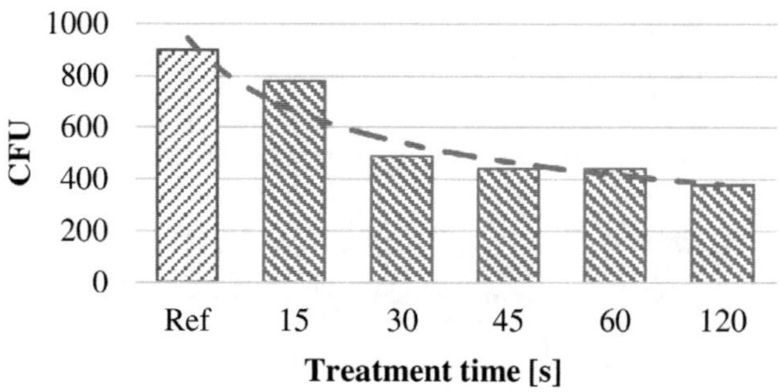

Figure 8. Bacteria inactivation rate after LARGE treatment as a function of treatment time. Parameters used: Ar-N_2 mixture (14 Vol.-%, i.e. 25 l/min + 5 l/min) and arc current of 50 A.

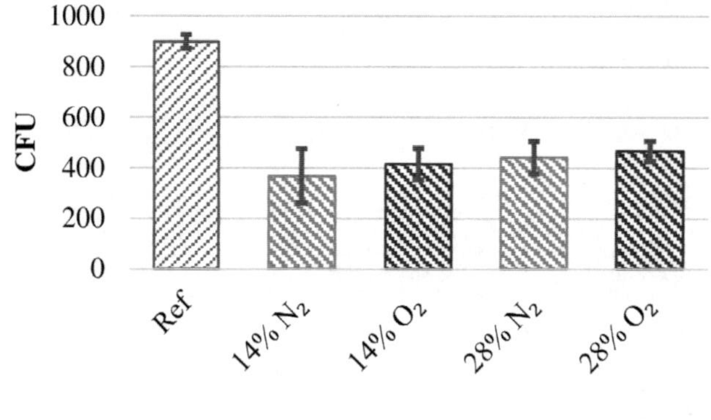

Figure 9. Influence of the plasma gas composition on the bacteria inactivation using thermal LARGE-plasma. Parameters used: arc current of 50 A and treatment time of 60 s.

5. Discussion

As could be shown, whether a thermal or a non-thermal generator is being used, the choice of plasma carrier gas seems to be important for the microbicidal effect of a plasma. Nonetheless, only with thermal plasma a significant inactivation of Geobac. stearothermophilus spores could be achieved. In addition to that, the best results could be achieved for an Ar-N_2 (14 Vol.-%) mixture. A possible reason for that could be an increased generation of RNS (and maybe ROS) due to secondary reactions with ambient air. A change in the pH-value of the treated TSA-dishes could also be a conceivable explanation. Interestingly, a similar addition of oxygen or an increased nitrogen concentration seems to provide slightly worse results. At the moment, no unambiguous explanation can be provided for this effect. Further investigation of different gas compositions coupled with spectroscopic measurements may help to understand this effect.

Moreover, a slight increase in arc current (from 40 A to 50 A) also improved the effects, the positive effect was however much lesser in comparison to gas composition change. As the intensity of UV radiation is expected to increase with rising current, the main killing agents seem to be RNS and ROS. Thermal degradation of the agar, which could be partially observed after treatments with the non-thermal plasma generator, could not be seen after LARGE treatments. Although the estimated energy densities being in a similar range, significantly lower gas temperatures when using the LARGE could be measured (LARGE: 23.6 °C; PB3: 84.1 °C) in the treatment distance of 60 mm. Thus, heat seems not to be responsible for the killing.

All in all, the reduction in disinfection rates of Geobac. endospores amount only to about 50 %, but keeping in mind the test conditions (static treatment, 60 mm distance), there is no doubt about that far better results are possible if the plasma generator is being moved or the sample is directly treated with plasma.

6. Conclusion and Outlook

As the investigation results show, thermal plasmas can be applied for large-area bacteria inactivation on temperature-sensitive surfaces. Due to the wide range of possible parameter adjustments, the long arc plasma generator LARGE showed good potential for sterilization applications. Furthermore, a direct sterilization of Geobac. stearothermophilus endospores was possible, which was not the case for the non-thermal plasma generator. A significant reduction of CFUs just after 30 s treatment time could be achieved – further improvements possible with detailed parameter adjustments (for ex. treatment distance) and other testing conditions (non-static treatment). Most importantly, no thermal load of the treated surfaces could be determined. The main killing agents, for now, seem to be the reactive nitrogen and oxygen species. Nevertheless, as discussed in chapter 5, there are still some open questions regarding the observed effects. The exact plasma agents produced by the thermal plasma generator and responsible for bacteria inactivation remain to be explored in future work.

References

[1] K-D Weltmann et al. 2015 Plasma sources for medical applications - a comparison of spot like plasmas and large area plasmas *ICRP-9/GEC-68/SPP-33 Conf. Proc.* Honolulu

[2] K G Kostov et al. 2015 Inactivation of Candida albicans by Cold Atmospheric Pressure Plasma Jet *IEEE Transactions on Plasma Science* vol **43** (3) pp 770-775

[3] Y H Kim et al. 2014 Measurement of reactive hydroxyl radical species inside the biosolutions during non-thermal atmospheric pressure plasma jet bombardment onto the solution *Plasma Chem. Plasma Process* vol **34** pp 457-472

[4] X Lu; T Ye; Y Cao et al. 2008 The roles of the various plasma agents in the inactivation of bacteria *J. Appl. Phys.* vol **104** (053309)

[5] M Laroussi and F Leipold 2004 Evaluation of the roles of reactive species, heat, and UV radiation in the inactivation of bacterial cells by air plasmas at atmospheric pressure *IJMS* **233** pp 81-86

[6] O Lunov, V Zablotskii, O Churpita et al. 2014 Cell death induced by ozone and various non-thermal plasmas: therapeutic perspectives and limitations *Scientific Reports* **4** (7129) DOI: 10.1038/srep07129

[7] J Heinlin, G Isbary, W Stolz et al. 2011 Plasma applications in medicine with a special focus on dermatology *J. of the European Academy of Dermatology and Venereology* vol **25** pp 1-11

[8] I Tanarro, V J Herrero, E Carrasco et al. 2011 Cold plasma chemistry and diagnostics *Vacuum* vol **85** pp 1120-1124

[9] Positionspapier des Nationalen Zentrum für Plasmamedizin e. V. 2014 Risikopotenzial und Anwendungsperspektiven von kaltem Atmosphärendruckplasma in der Medizin *Plasmakurier* vol **1** pp 42-51

[10] H C Baxter, G A Campbell, P R Richardson et al.2006 Surgical Instrument Decontamination: Efficacy of Introducing an Argon: Oxygen RF Gas-Plasma Cleaning Step as Part of the Cleaning Cycle for Stainless Steel Instruments *IEEE Transactions on Plasma Science* vol **34** (4) pp 1337-1344

[11] M Y Alkawareek, Q Th Algwari, G Laverty et al. 2012 Eradication of Pseudomonas aeruginosa Biofilms by Atmospheric Pressure Non-Thermal Plasma *PLOS ONE* vol **7** (8) e44289

[12] X Liu, F Hong, Y Guo, J Zhang, J Shi 2013 Sterilization of Staphylococcus Aureus by an atmospheric non-thermal plasma jet *Plasma Sci. and Technol.* vol **15** (5) pp 439-442

[13] G Isbary et al. 2010 A first prospective randomized controlled trial to decrease bacterial load using cold atmospheric argon plasma on chronic wounds in patients *British J. of Dermatology* vol **163** (1) pp 78-82

[14] S Arndt, P Unger et al. 2013 Cold Atmospheric Plasma (CAP) Changes Gene Expression of Key Molecules of the Wound Healing Machinery and Improves Wound Healing In Vitro and In Vivo *PLOS ONE* vol **8** (11) e79325

[15] A Gutsol, N D Vaze, K P Arjunan et al. 2008 Plasma for Air and Water Sterilization, Plasma Assisted Decontamination of Biological and Chemical Agents *NATO Science for Peace and Security Series*

Series A: Chemistry and Biology pp 21-39

[16] A Mizuno 2007 Industrial applications of atmospheric non-thermal plasma in environmental remediation, *Plasma Phys. Control. Fusion* vol **49** pp A1-A15

[17] L Zhanguo, L Ying, C Peng, Z Hongjie 2013 Surface Decontamination of Chemical Agent Surrogates Using an Atmospheric Pressure Air Flow Plasma Jet *Plasma Sci. Technol.* vol **15** (7)

[18] D B Kim, B Gweon, S Y Moon, W Choe 2009 Decontamination of the chemical warfare agent simulant dimethyl methylphosphonate by means of large-area low-temperature atmospheric pressure plasma, *Curr. Appl. Phys.* vol **9** pp 1093-1096

[19] Th Von Woedtke, H-R Metelmann, J Lademann et al. 2015 Clinical application of cold atmospheric plasma – state and perspectives *Conf. Proc. 22nd ISPC* Antwerp Belgium

[20] Y Ma, G J Zhang, X-M Shi et al. 2008 Chemical mechanisms of bacterial inactivation using dielectric barrier discharge plasma in atmospheric air, *IEEE Transactions on Plasma Science* vol **36** (4) pp 1615-1620

[21] Z Cao, Q Nie, D L Bayliss et al. 2010 Spatially extended atmospheric plasma arrays *Plasma Sources Sci. Technol.* vol **19** (025003)

[22] Y Song, D Liu, L Ji et al. 2012 Plasma Inactivation of Candida albicans by an Atmospheric Cold Plas-ma Brush Composed of Hollow Fibers, *IEEE Transactions on Plasma Science* vol **40** (4) pp 1098-1102

[23] L C Farrar, D P Haack et al. 2000 Rapid Decontamination of Large Surface Areas *IEEE Transactions on Plasma Science* vol **28** (1)

[24] R Hartmann 2000 Entwicklung eines Breitstrahl-Plasmabrenners zur Behandlung planarer Oberflächen *Doctoral thesis* Universität der Bundeswehr München

[25] H-M Kim, M Dzulko, E Theophile, K Landes 2003 Aktueller Entwicklungsstand des Breitstrahl-Plasmagenerators LARGE und dessen Anwendungsmöglichkeiten, *5. Industriefachtagung Oberflächen- und Wärmebehandlungstechnik,* TU Chemnitz

[26] K D Landes, M Dzulko, E Theophile, J Zierhut 2002 New developments in DC-plasma torches *High Temperature Material Processes* vol **6** (3)

[27] Information materials published by Relyon Plasma GmbH, Regensburg, Germany https://www.relyon-plasma.com/applications/sterilization/?lang=en

[28] A D Morris, G B McCombs et al. 2009 Cold plasma technology: bactericidal effects on Geobacillus stearothermophilus and Bacillus cereus microorganisms *J. of Dental Hygene* vol **83** (2) pp 55-61

[29] C A Batt and M L Tortorello 2014 *Encyclopedia of Food Microbiology, Second Edition* vol 1

[30] O Yardimci and P Setlow 2010 Plasma Sterilization: Opportunities and Microbial Assessment Strategies in Medical Device Manufacturing *IEEE Transactions on Plasma Science* vol **38** (4) pp 973-981

[31] J Jeon 2014 Surface Micro-Discharge (SMD) – Analysis of the Antimicrobial Effect and the Plasma Chemistry *Doctoral thesis* Ludwig-Maximillians-Universität München

IOP Conf. Series: Journal of Physics: Conf. Series **825** (2017) 012018 doi:10.1088/1742-6596/825/1/012018

Anode energy transfer in a transient arc

F Valensi[1*], P Ratovoson[1], M Razafinimanana[1] and A Gleizes[1,2]
[1] *LAPLACE (Laboratoire Plasma et Conversion d'Energie), CNRS-INPT-Université Toulouse III, 118 Route de Narbonne, 31062 Toulouse cedex 9, France*
[2]*CNRS, LAPLACE, 118 route de Narbonne, 31062 Toulouse Cedex 9*

E-mail : flavien.valensi@laplace.univ-tlse.fr

Abstract. This work deals with experimental investigation of a transient arc. Arc configuration and electrode erosion were studied in order to quantify the energy transfer to the electrodes as a function of maximal current, time constant and electrodes material. Experiments with two consecutive arcs allow demonstrating non stationary behaviour of the arc electrode interaction. This is due to the fact that while the duration of the experiments is far larger than plasma phenomena time constants, it is comparable to those of electrode heating and melting processes.

1. Introduction

Transient arc can occur during the separation of contacts in an electric circuit. They are characterized by a short duration of a few milliseconds to a few hundreds of milliseconds. The current can reach high value, usually at least a few hundred or a few thousand amperes. Such arcs occur in circuit breakers operation [1] but can also be observed in rail transportation systems when the contact is lost between the pantograph and the overhead line [2]. In this case the arc can lead to significant degradation of the electrode material (particularly the contact strip of the pantograph) in addition to abrasive wear. The distance between the electrodes can increase from near contact to a few millimeters or even a few centimeters. This means that the voltage drop of the arc is at least a few tens of volts and the resulting power is high enough to cause strong electrode erosion. Both electrodes are degraded but the damages are more important at the anode. It is then important to get a better understanding of the involved phenomena in order to further improve the systems performances. In particular a lot of efforts are made to increase the lifetime of the electrodes contacts by developing new materials. The work reported here is dedicated to the study of electrode erosion for graphite and copper contacts. Samples made of pure materials have been tested as well as a two components electrodes made of graphite loaded with 25% Cu. The general arc behavior is first presented for several arc conditions (current maximal value, duration of the pulse and electrode gap). Then the anode erosion is studied for the three materials and the results are discussed in term of energy.

2. Experimental setup

The arc is generated using a capacitor bank discharge. The maximal current and the time constant of the RC circuit can be set independently; detailed description of the setup is given in [3]. The arc discharge is triggered by the closure of a thyristor: the current rises up to its maximal value in less than one millisecond, according to the thyristor switching on characteristic. Indeed, the VSKT162/12PBB modules used for this setup can reach their nominal continuous current (160 A) in a few microseconds according to the datasheet but the current rise slope decreases for high current in pulsed mode operation. The arc initiation can be performed with a small copper wire (for small electrode gap) or without contact between the electrodes by applying a high voltage pulse when the thyristor closing is triggered. The current then decreases following quite closely the exponential decay of a RC circuit.

The arc phase ends when the voltage becomes too low but it is also possible to shut down the arc after a given delay with a time step of 1 ms. However due to the thyristor technology the switching off is not instantaneous and arc actual extinction occurs a few milliseconds later. The capacitor bank can also be split in two independent parts, thus allowing double shots separated by a few milliseconds. Experiments were performed with peak current up to 2 kA and time constants from 24 to 92 ms. In the case of double arcs the delay was set to 12 ms.

The cathode is made of a 6 mm cylindrical graphite rod with tapered end. Two anode diameters (6 mm and 15 mm) were used in the case of copper; tests with graphite were performed with a 10 mm diameter anode. The square contact strip samples (25 wt.% Cu) were 15 mm wide. In all experiments the anode thickness was about 10 mm. The height of the cathode tip above the anode (h) can be set from 1 to 10 mm. The arc geometric configuration and electrode erosion are observed with high speed camera (Phantom V9) using acquisition speed from 4000 to 6000 fps. Two filters were used, one centred at 515 nm (FWHM: 20 nm), corresponding to Cu I emission lines and one centred at 750 nm (FWHM: 25 nm), corresponding to N I emission lines. This allows to get qualitative information about plasma composition and the repartition of elements coming from the anode (copper) or already present in the gas (nitrogen). The mass loss is measured by weighting the samples before and after each experiment. In addition the electrode ablation can be monitored during the arc through high speed imaging and the surface of the anode samples are observed after arc extinction. The electrical parameters (arc current I and voltage U) were also recorded with a time resolution of 25 μs.

3. Results and discussion

3.1. Arc shape and electrode erosion observation

The typical shape of the arc corresponds to a cone with the tip at the cathode for all studied anode materials. An example is shown in figure 1 for a 6 mm diameter copper anode with three values of maximal current and a time constant of 24 ms. This behavior can be attributed to Maecker effect, as a result of the electromagnetic force producing a pinch at the cathode tip [4]. Then the arc behaves more like a constricted arc than a free burning arc, which is consistent with results found in literature for short length (< 1 cm) and high current (> 100 A) arcs [5]. The comparison of images obtained with interference filters centered on copper and nitrogen lines provides qualitative information on these elements repartition. While nitrogen is distributed rather homogeneously in the plasma, a darker zone (indicating a low neutral nitrogen concentration area) can always be observed below the cathode for imaged recorded with the filter centered on nitrogen lines [3]. This can be explained by the prevalence of ionized lines over neutral lines considering the high temperature (> 20 000 K) [5,6] encountered. Its size differs according to the nature of the anode material: it is larger and can be observed for a longer time with pure graphite than with copper containing anode. This effect can be attributed to the higher plasma cooling due to strong copper vapor concentration and consequently increased radiation losses.

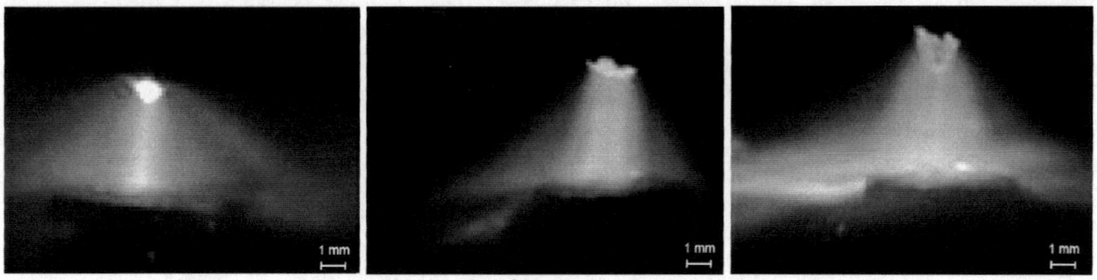

a) b) c)

Figure 1. Arc shape (λ = 515 nm) 0.5 ms after ignition; 6 mm copper anode, h= 4mm, τ = 24 ms: a) I_{max} = 500 A, b) I_{max} = 1000 A and c) I_{max} = 2000 A.

Some information about electrode erosion can be obtained from the high speed image recording. In the case of pure graphite anodes the mass loss is limited and the erosion seems to be a continuous process. This is consistent with the fact that carbon directly sublimates and there is no molten phase. When the

anode is constituted by a mixture of carbon and copper the erosion is slightly higher, but the main difference is that one can observe small particles ejected from the surface of the anode. In contrast the erosion of pure copper anodes is much stronger and large deformation of the electrode can be observed. Metal droplets are ejected from the arc attachment area and in the case of 6 mm diameter anodes, melted copper can flow down from the surface.
While the current decreases along the discharge following an exponential decay, the maximal erosion is not observed at the beginning of the arc. In the first step, the presence of copper vapor is weak and no electrode deformation is visible. Then the erosion increases and the arc radiation is dominated by metallic lines while the melting of the anode surface becomes visible. Finally, at the end of the arc, radiation from nitrogen is dominant again and no more anode erosion occurs. The time needed to reach significant erosion and the duration of the highest erosion step as a function of operating parameters is reported in table 1.

Table 1. Duration of the steps associated to 6 mm diameter copper anode erosion (h = 4 mm).

Maximal current I_{max} (A)	Time constant τ (ms)	Erosion beginning (ms)	Maximal erosion (ms)
500	24	3	6-8
	41	3	8-12
	91	3	40-70
1000	24	2	8-16
	41	2	8-20
2000	24	1.25	8-16

It appears that the first step depends mainly on the maximal current but the time constant of the discharge has no influence on its duration. On the other hand, the time needed to reach maximal erosion depends on both parameters. However at high current (1000 and 2000 A) there is no more effect of current increase which can be attributed to minimal time needed to heat up and melt the electrode. When a 15 mm diameter anode is used, the deformation of the electrode is reduced (no molten metal flowing from the electrode) and the erosion is less irregular. The duration of the first step remains unchanged (for instance, 2 ms for a 1000 A current) but the time needed to reach maximal erosion is shorter (between 6 and 10 ms). The duration of the observable erosion step is also reduced, which can be explained by the fact that the heat flux is transferred on a larger volume in the case of larger anode samples. The power needed to heat and melt the electrode is greater and the time during which erosion is possible is reduced.

3.2. Anode ablation measurement

3.2.1. Ablated mass. We have shown in a previous work [3] that the mass loss increases with current and time constant is more than ten times higher with copper than with graphite for a 6 mm diameter anode. It is negligible in the case of graphite for a current of 500 A and reaches 4.8 mg ± 5% for $\tau = 24$ ms and $I_{max} = 2000$ A. Under the same conditions it is up to 150 mg ± 9% for copper. These results are consistent with the fact that graphite appears to be ablated only through sublimation (at 4100 K) while copper melts at 1357 K. While copper vaporization also occurs, melted metal can also leave the anode as ejected droplets which significantly increases mass loss in the case of small diameters anodes.

3.2.2. Parametric study. The first studied parameter is the anode diameter. In the case of graphite (where the results are not influenced by electrode melting) the results for a 1000 A arc are given in table 2. The last column shows the total electrical energy for the test. It appears that the ablation decreases when diameter increases as reported by Heberlein et al. [7]. For short length arcs and flat anodes, arc attachment is quite diffuse as one can see in figure 1. However the anodic root expansion is limited due to magnetic pinch effect and thermal and electrical conductivity effects in the solid metal. For low current, the anodic root dimension is smaller than the anode diameter and very low ablation occurs, especially with refractory material such as graphite. Energy loss through conduction and convection are dominant and few energy is available for electrode erosion. The latter then increases with current. When it is high enough so that the anodic root reaches the size of the anode, further increase will lead to higher current density, especially on the edge of the electrode. The power transferred to the anode and the resulting ablation will then significantly increase with current.

Table 2. Ablated mass for I_{max}=1000A (τ= 41ms), h = 4 mm.

Material	ϕ (mm)	Ablated mass (mg)	Uncertainty (%)	Energy (J)
Graphite	6	2.46	5.03	1223
Graphite	10	1.08	4	1158
Copper	6	53	10.6	1091
Copper	15	6.3	8.9	998

This explains why for a given current the anode ablation is weaker when the diameter is larger. Besides, the material volume which must be heated before erosion occurs is larger. In the case of graphite, our experimental conditions correspond to the low erosion mode, below the current limit yielding to higher erosion. On the opposite, there is a strong difference between the two tested diameters in the case of copper. Results tend to show that the anodic root is smaller than 15 mm and in this case the ablation remains weak.

The second studied parameter is the arc duration. It depends on the time constant but the possibility to cause arc extinction at a chosen time also allows studying the evolution of erosion over time. The results for copper anodes with diameter of 6 and 15 mm and an arc length of 2 mm are shown in figure 2. It appears that erosion increases over time during about 50 ms before it reaches a plateau (figure 2a). This confirms the fact that below a given value of current the erosion becomes negligible: at this time the intensity of the current becomes lower than 300 A. The behavior is similar for the 15 mm diameter electrode (figure 2b) but the increasing erosion step is much shorter. Strong erosion occurs only during the first 5 ms.

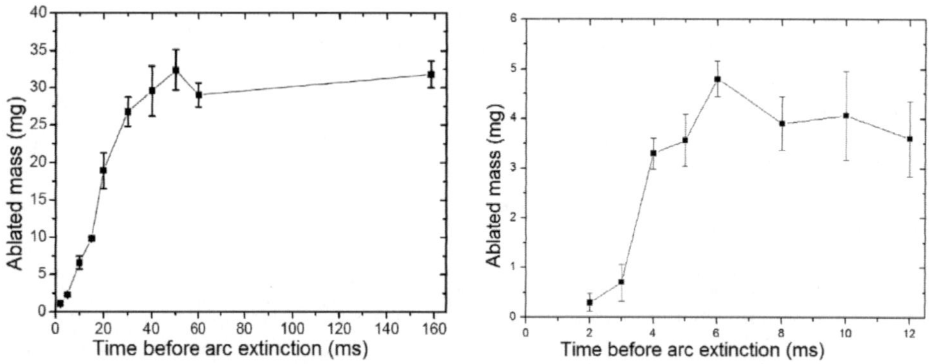

a) b)

Figure 2. Ablated mass as a function of arc duration for h = 2 mm, I_{max}=1000A; a) anode diameter of 6 mm and τ= 41 ms, b) anode diameter of 15 mm, and τ= 27 ms

The high erosion step duration appears to be different when the arc length changes from 4 to 2 mm. In order to further investigate, this evolution of ablated mass has been studied as a function of the arc length. The peak currents set at 1000 A and a time constant of 41 ms for copper anodes. The results for the two tested anode diameters (6 and 15 mm) are presented in figure 3. In the case of the smaller diameter the influence of arc lengths seems to be low but results show clearly an increase of mass loss from 2 to 4 mm. While for the larger anode diameter, the ablation is lower the behavior is the opposite. There is a strong decrease in ablation when arc length increases, even if for distance greater than 8 mm

IOP Conf. Series: Journal of Physics: Conf. Series **825** (2017) 012018 doi:10.1088/1742-6596/825/1/012018

there is also the beginning of a plateau. Then for a given arc root extension, the size of the anode seems to have determinant influence on the dependence of arc length.

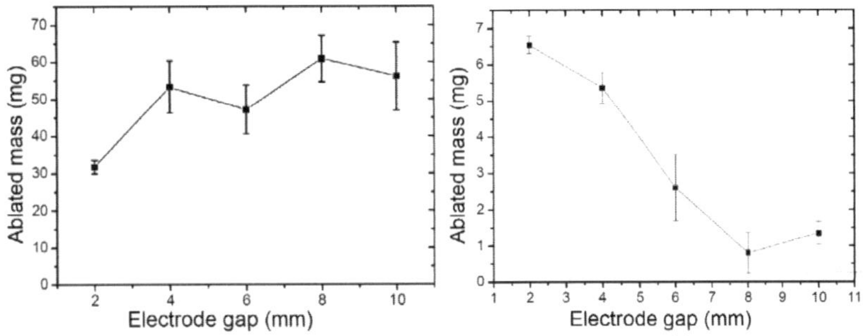

a) b)

Figure 3. Ablated mass as a function of electrode gap h for I_{max}=1000A and τ= 41ms; a) copper anode diameter of 6 mm, b) copper anode diameter of 15 mm

3.2.3. Comparison with literature. In the case of copper anode, our work can be compared to those of Devautour *et al.* [8] with arc duration between 5 and 20 ms, current ranging from 200 and 1000 A and electrode diameter of 8 mm. The ablated mass at anode increases from 1 to 2 mg for maximal current of 400 and 600 A, respectively and reaches 6 mg at 1000 A. These results are comparable to ours for the 6 mm diameter anode for the two lowest values and for the 15 mm anode for the highest. This corresponds to conditions where arc attachment is smaller than the electrode diameter. The influence of arc length in the case of carbon anodes has been studied by Mentel *et al.* [9]: for electrode gap which does not exceed 1 cm, the electrode ablation increases as a function of arc length. The correlation between energy transfer and electrode erosion can be considered by relating ablation to current intensity (or electrical power) at a given instant or to the energy for a given period. When the current intensity is not constant over time the ablation can also be related to electric charge [10]. For a graphite anode, our results show that the ablation is negligible for current below 500 A. Then, one can consider the total electric charge Q_{tot} (integration over all the pulse duration) and the partial charge Q_{par} corresponding to the time during which current is greater than 500 A. We have reported the values we get using this method in table 3.

Table 3. Ablated mass and electric charge for the graphite anode.

I_{max} (A)	τ(ms)	ϕ(mm)	Q_{tot}(C)	Q_{par} (C)	m (mg)	m/Q_{tot} (mg/C)	m/ Q_{par} (mg/C)
2000	24	6	46,8	35,5	4,78	0,102	0,134
1000	24	6	20,6	10,84	1,23	0,06	0,114
	41	6	34,8	18	2,46	0,07	0,137
		10	34,8	18	1,08	0,031	0,06

When considering the calculation for the entire pulse, our results are in agreement with those of Lefort *et al.* [10] experimentally obtained with a current of 70 A (0.05 mg/C). The values they calculated for currents of 1000 A and more (0,16 mg/C) are also compatible with that we obtained with small diameter electrodes when considering only time during which current is above 500 A.

3.3. Energy transfer at the anode

3.3.1. Arc total power and energy. The instantaneous total power injected in the arc can be calculated from instantaneous current and arc voltage:

$$p(t) = u_{arc}(t) \times i(t) \tag{1}$$

with

$$u_{arc} = u_{col}(t) + u_{elec}(t) \quad (2)$$

In equation (2), $u_{col}(t)$ corresponds to the arc column voltage and $u_{elec}(t)$ the electrodes voltage drop. The power temporal evolution for three currents is shown in figure 4. It appears that the power is always smaller with copper, which can be attributed to smaller voltage drop and the strong presence of metal vapors leading to decrease the arc voltage with this material [11].

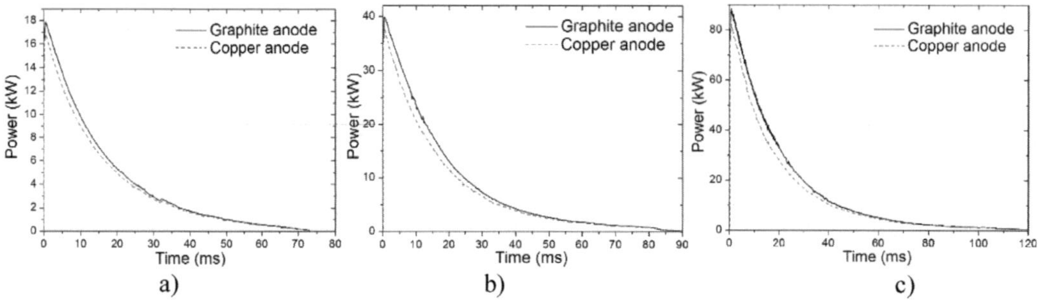

Figure 4. Total power injected in the arc for copper and graphite anodes, τ = 24 ms: a) I_{max} = 500 A; b) I_{max} = 1000 A; c) I_{max} = 2000 A.

The total energy can be calculated by integrating instantaneous power over the total discharge time. Results are presented in table 4 for three currents, three time constants and both anode material. For each case, according to results above, the arc energy is greater with graphite than with copper.

Table 4. Total injected energy for graphite and copper anodes.

I_{max} (A)	Time constant τ (ms)	Total injected energy E (J)	
		Graphite	Copper
500	24	298	275
	41	522	479
	91	1238	1083
1000	24	731	651
	41	1223	1091
2000	24	1823	1595

3.3.2. Estimation of anode energy transfer. The energy transferred to the anode (for a 200 A arc [12]) corresponds mainly to about 70 % of the quantity $(V_a + \phi_s)\times j_e$, where V_a is the anode voltage fall, ϕ_s the material work function and j_e the current density. When integrating over the anode surface, it yields to $(V_a + \phi_s)\times I$. The work function for copper and graphite is 4.5 and 4.9 eV, respectively. Higher uncertainties are associated to anode voltage fall and various values can be found in literature ([5], [13]). On the basis of our previous results [11] the voltage can be estimated to a few volts (at least 1 V). Considering that the power transferred to the anode P_{anode} is proportional to current it can be written as:

$$P_{anode} = I.\ U_{anode} \quad (3)$$

With the equivalent voltage V_{anode} equals to 9.5 V ± 1.5 V. This represents 20% to 30 % of total voltage for arc length of 1 cm to a few mm. Then one can consider that the energy transferred to the anode can reach one third of arc total energy in our experimental conditions.

3.3.3. Correlation between energy transfer and ablation – arc length influence. In order to consider the influence of deposited energy on material ablation, the most relevant criterion is the power density. According to results presented in figure 3, two different anode diameters lead to different values and evolution as a function of arc length for ablated mass. When electrode gap increases, the voltage and the electric power increase. This leads to higher arc radiation loss that contributes to heat up the anode, (even if in our conditions radiation is not the main contribution of the energy balance). For a longer arc, the anodic root size increases and the power density at the anode is lower which means less ablation. This occurs for arc shorter than a few cm, for greater length the anodic root is constricted. Two plasma jets from both electrodes join in a diffuse intermediary region [14]. Considering that the anode voltage fall remains constant when arc length increases, it is then possible to explain the contradictory behavior observed in figures 3a and 3b. For small anode diameter, the anodic root covers all anode surface and when arc length increases the ablation moderately increases due to higher radiation transfer from the arc. For greater anode diameter, the arc root (initially smaller than anode diameter) increases with arc length, which decreases the power density and then the ablated mass.

3.4. Copper anode vaporization

3.4.1. Procedure for estimation of the vaporized mass. The mass loss that is obtained from weighting measurements corresponds to melted (and ejected as droplets) and vaporized metal. The high speed imaging of the arc has shown that copper vapors are present during all the discharge, even if the amount is very low for the first milliseconds [3]. The mass of metal required to get visible copper vapor is very low (0.1 mg for a 100 ms arc) when compared to the measured ablated mass (1 to 150 mg).

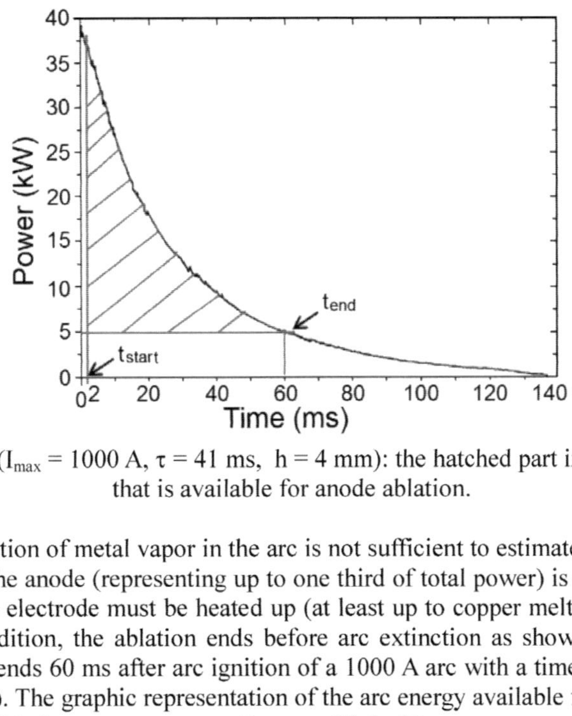

Figure 5. Arc power (I_{max} = 1000 A, τ = 41 ms, h = 4 mm): the hatched part indicates the arc energy that is available for anode ablation.

Then, qualitative detection of metal vapor in the arc is not sufficient to estimate the vaporized part. All energy transferred to the anode (representing up to one third of total power) is not directly responsible of mass loss, since the electrode must be heated up (at least up to copper melting temperature) before ablation occurs. In addition, the ablation ends before arc extinction as shown in figure 2. One can consider that ablation ends 60 ms after arc ignition of a 1000 A arc with a time constant of 41 ms (the plateau starts at 50 ms). The graphic representation of the arc energy available for ablation is presented in figure 5, the actual ablation energy representing one third of it.
The start and end of ablation are indicated as t_{start} and t_{end}. The corresponding minimal power needed for electrode erosion is 5 kW. These two instants cannot be evaluated with precision and while the relative uncertainty over t_{stard} has low consequences (t_{start} ranges from 1 to 3 ms) the appreciation of t_{end} is more critical. While the mass loss measurements as a function of time (figure 2) lead to a value of t_{end} between 40 and 60 ms, high speed observation tends to show anode erosion up to 80 ms. In order to get a parameter independent of time constant and arc duration, the current corresponding to time of ablation end has been considered. This intensity ranges from 200 and 400 A for the three studied currents and time constants. The energy responsible for anode ablation is then obtained by integrating the power according to these limits (hatched part of figure 5) and multiplying the result by 0.3. Results are presented in table 5.

Table 5. Estimation of the energy needed (in J) for copper anode ablation (6 mm diameter, h = 4 mm)

Energy (J)	I_{max}=500A			I_{max}=1000A		I_{max}=2000A
$I_{end\ of\ erosion}$ (A)	τ=24ms	τ=41ms	τ=91ms	τ=24ms	τ=41ms	τ=24ms
400A	2.13	4.78	10.59	53.46	89.34	241.02
316A	7.68	15.24	34.29	69.9	118.92	275.07
300A	9.27	16.83	37.8	73.35	124.05	283.86
200A	20.52	37.62	84.33	97.74	167.58	328.17

These values of energy have to be interpreted considering the energy needed to heat and melt solid copper then to heat and vaporize liquid copper. For a mass of 1 g it corresponds to 583 J (when starting at room temperature) and 5454 J, respectively.

3.4.2. Discussion. On the basis of high speed imaging, one can consider that for current below 500 A (where ablation is low) the anode surface presents a small melted pool with low perturbation and that copper ablation is mainly due to vaporization (no ejected droplets). Then, for a current of 500 A the energy available for anode erosion (presented in table 5) is roughly equal to the energy needed to vaporize the ablated mass (table 6). This corresponds to a current threshold between 316 and 363 A, which is in agreement with the results of ablation measurement as a function of time presented in figure 2. Then a value of 340 ± 24 A can be considered. The melted and vaporized part of copper can then be calculated from the thermophysical data indicated above, the energy available and the ablated mass. The results for the three currents and time constants are given in table 6.

Table 6. Estimation of the melted and vaporized quantities of copper for a 6 mm diameter anode.

I_{max} (A)	Time constant τ (ms)	Ablated mass (mg)	Energy needed for total ablated mass vaporization E_{vap} (J)	Estimated melted mass (mg)	Estimated vaporized mass (mg)
	24	1.05	6.4	0	1.05
500	41	1.72	10.5	0	1.72
	91	2.78	17	0	2.78
1000	24	14.81	90.7	10.1	4.71
	41	53.03	324.8	13.65	39.38
2000	24	150.6	922	29.2	121.4

For current below 500 A, the ablated mass corresponds mainly to vaporized matter. When current increases the contribution from metal that is melted and ejected (as droplets) from the anode is greater. For the highest intensities, the melted part becomes dominant when compared to vaporized part. These conclusions are valid for small diameter anodes, where the arc anode root covers the entire electrode surface event at low current.

3.5. Application to the carbon-copper electrode

In addition to pure graphite or copper electrodes, experiments were made with contact strop samples that contain 75 wt.% carbon and 25wt.% copper. At low current (500 A and τ = 91 ms) the mass loss (3 ± 0.6 mg) is similar to that observed with copper anodes. For higher current the ablation does not increase as much as for copper but is greater than for graphite. For instance with a maximal current of 1000 A (and a time constant of 41 ms) the mass losses are 2.5 ± 0.25 mg, 6.3± 0.57 mg and 4.1± 0.19 mg for graphite, copper and graphite-copper anode samples, respectively. Then there is no simple relation between the composition and the erosion resistance of the material.

As in the case of copper or graphite, the influence of ablated mass as a function of arc length (electrode gap) has been studied. The results for two currents and time constants (I_{max} = 500 A, τ = 91 ms and I_{max} = 1000 A, τ = 41 ms) are presented in figure 6. While the anode size is larger than the arc root the ablation increases with arc length, contrarily to copper with identical conditions. High speed imaging shows that in this case arc root size decreases for greater arc length, leading to higher power density and higher erosion. In addition, one has to consider that copper and graphite anodes are solid homogeneous materials while the graphite-copper samples are produced through sintering, resulting in non-homogeneous materials.

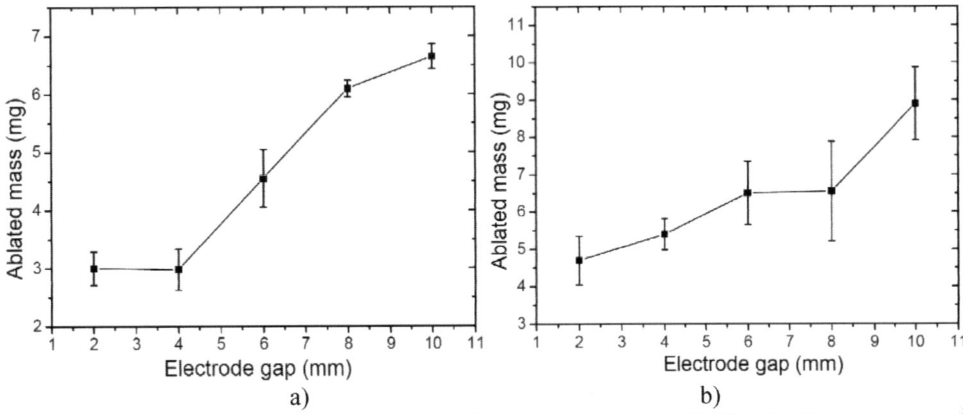

Figure 6. Ablated mass as a function of electrode gap h for C+25wt.% Cu anodes a) I_{max}=500A and τ= 91ms; b) I_{max}=1000A and τ= 41ms;

The influence of the electrode temperature evolution has been studied through multiple successive arc experiments. The results have been obtained with one arc with extinction triggered after 6 ms and two 3 ms arcs separated by 9 ms delay. While the duration of the high current step is the same for both experiments, the ablation appears to be smaller in the second case: it drops from 4.8 ± 0.29 mg to 2.9 ± 0.21 mg in the case of pure copper. The trend is similar for the graphite-carbon mixture (from 1.4 ± 0.07 mg to 1.1 ± 0.08 mg) even if the difference is smaller. These results can be related to the heating process of the electrode. When the arc is extinguished after 3 ms the material starts cooling down and needs to be heated up again during the beginning of the second pulse. Then, even if the total time for high power is the same as for the 6 ms arc, less energy is available to cause material ablation. This proves that the anode temperature evolution is relevant considering the arc time scale: the time needed to heat up the electrode is not negligible when compared to the duration of the studied arcs. A duration of 9 ms is sufficient to lead to significant electrode cooling. This justifies that the non-stationary aspect of the plasma-electrode interaction has to be considered.

4. Conclusion

The results show that the ablation in the case of copper is much stronger than in the case of graphite. The influence of geometric (anode size and arc length) and electric (maximal current and pulse duration) parameters has been studied for both materials. For a given anode sample, there is a current intensity threshold (320 A for copper and 500 A for graphite) below which no significant erosion occurs. The size of the anodic root has been found to be determinant on ablation mechanism. The repartition between melted and vaporized metal leaving the anode is also dependent on current: at low intensity (500 A) vaporization is dominant while at higher current intensities (1000 and 2000 A) ejection of melted metal droplets becomes the main matter loss mechanism. The erosion for a mixture of graphite and copper is higher than with pure graphite, but lower than with pure copper. This composition (25 wt.% Cu) seems to correspond to a good compromise between electrical conductivity (higher for copper) and material ablation resistance (higher for graphite). Besides it is known that friction wear resistance is higher with large graphite content.

Acknowledgements: The authors wish to thank Pr. A. Mpanda from ESIEE (14 quai de la Somme, BP10100 - 80082 Amiens Cedex 2, France.) for the contact strip samples.

References

[1] McBride J W and Weaver P M 2001 Review of arcing phenomena in low voltage current limiting circuit breakers *IEE Proc.-Sci. Meas. Tech* **148**-1 pp 1–7

[2] Bormann D, Midya S and Thottappillil R 2007 DC components in pantograph arcing: Mechanisms and influence of various parameters *Proceedings of 18th International Zurich Symposium on Electromagnetic Compatibility, Munich, Germany* pp 369–372

[3] Ratovoson P, Valensi F, Razafinimanana M and Tmenova T 2014 Spectroscopic study and high speed imaging of a transient arc *Journal of Physics: Conference Series* **550** 012012

[4] Finkelnburg W and Maecker H 1956 Electric Arcs and Thermal Plasmas In: *Encyclopedia of Physics*, vol. XXII, Springer-Verlag, Berlin

[5] Hsu K C, Etemadi K and Pfender E 1983 Study of the free-burning high-intensity argon Arc. *J. Appl. Phys*. **54** pp 1293-1301

[6] Freton P, Gonzalez JJ, Camy F, Peyret and Gleizes A 2003 Complementary experimental and theoretical approaches to the determination of the plasma characteristics in a cutting plasma torch. *J. Phys. D : Appl. Phys.* **36**, pp 1269-1283

[7] Heberlein J, Mentel J and Pfender E 2010 The anode region of electric arcs: a survey. *J. Phys. D: Appl. Phys.* **43** 023001 (31p)

[8] Devautour J, Chabrerie J P and Testé Ph 1993 The study of thermal-Processes in an electrode submitted to an electric-arc *J. Physique III* **3** pp 1157-1166

[9] Mentel J and Heberlein J 2010 The anode region of low current arcs in high intensity discharge lamps. *J. Phys. D: Appl. Phys.* **43** 023002 (20p)

[10] Lefort A, Parizet MJ, El-Fassi SE and Abbaoui M 1993 Erosion of Graphite-electrodes *J. Phys. D: Appl. Phys.* **26** pp 1239-1243,

[11] Valensi F, Ratovoson L, Razafinimanana M, Masquère M, Freton P and Gleizes A 2016 Electrode voltage fall and total voltage of a transient arc, *J. Phys. D: Appl. Phys.* **49** 255202 (13p)

[12] Lago F 2004 In: *Modélisation de l'interaction entre un arc électrique et une surface: application au foudroiement d'un aéronef,* Ph.D Dissertation, Université Paul Sabatier, Toulouse III

[13] Lowke J J, Morrow R, Zhu P, Haidar J, Farmer A J D and Haddad G N 1992 The physics of free burning arc and their electrodes *Journal of High temperature chem. Processes* pp 549-556

[14] Bauchire J M, Hong D, Rabat H and Riquel G 2012 Radiation of transient high-current arcs: energy measurements in the optical range *Journal of Physics: Conference Series* **406** 012030